Lecture Notes in Mathematics

Edited by A. Dold and B. Eckmann

1003

Jean Schmets

Spaces of Vector-Valued Continuous Functions

Springer-Verlag
Berlin Heidelberg New York Tokyo 1983

Author

Jean Schmets
Institut de Mathématique, Université de Liège
15, avenue des Tilleuls, 4000 Liège, Belgique

AMS Subject Classifications (1980): 46 A 05, 46 E 10

ISBN 3-540-12327-X Springer-Verlag Berlin Heidelberg New York Tokyo
ISBN 0-387-12327-X Springer-Verlag New York Heidelberg Berlin Tokyo

Printed in Germany

Printing and binding: Beltz Offsetdruck, Hemsbach/Bergstr.
2146/3140-543210

INTRODUCTION

Let X be a Hausdorff completely regular space and E be a Hausdorff locally convex topological vector space. Then C(X;E) denotes the linear space of the continuous functions on X with values in E; in the scalar case (i.e. if E is $\mathbb{R}$ or $\mathbb{C}$) we simply write C(X).

The purpose of these notes is to characterize locally convex properties of C(X;E) by means of topological properties of X and of locally convex properties of E.

The scalar case has already been developed in [20] and it is quite convenient to recall briefly its contents. Its chapter I deals with the ultrabornological (resp. bornological; barrelled; quasi-barrelled) space associated to E. One finds in its chapter II a description of the realcompactification υX of X as well as the definition of the space $C_p(X)$, i.e. C(X) endowed with the most general locally convex topology of uniform convergence on subsets of υX. In its chapter III, the ultrabornological (resp. bornological; barrelled; quasi-barrelled) space associated to $C_p(X)$ is characterized. This gives of course a necessary and sufficient condition for $C_p(X)$ to have that property. Its chapter IV concerns conditions of separability or of weak-compactness in $C_p(X)$. Finally its chapter V gives an attempt to study the vector-valued case : it deals with the space $C_s(X;E)$, i.e. C(X;E) endowed with the simple or pointwise topology.

Since the publication of [20], its chapters I to IV have got few complementary results. It is not the case of its chapter V to which significant results have been added. This is due mostly to A. Defant, W. Govaerts, R. Hollstein, J. Mendoza Casas, J. Mujica, myself, ... Now it is possible to say that large parts of the study of the $C_p(X;E)$ spaces are settled.

The aim of these notes is to present these new results as a complement to the chapters I to IV of [20].

The chapter I contains a description of the spaces $C_p(X;E)$, i.e. C(X;E) endowed with the most general locally convex topology of uniform convergence on subsets of υX. The notion of the support of an absolutely convex subset of $C(X;E)$ is studied in the chapter II. The dual of $C_p(X;E)$ is characterized in the chapter III. The chapter IV gives most of the known results dealing with the $C_p(X;E)$ spaces and the ultrabornological, bornological, barrelled, quasi-barrelled and (DF) properties. Finally a special link in between the bounded and the sequentially continuous linear functionals on $C_p(X;E)$ is introduced in the chapter V.

It is a pleasure to thank Mrs. V. Berwart-Verbruggen who has done the typing very carefully. I am grateful to the editors for accepting these notes for publication in the Lecture Notes in Mathematics.

J. SCHMETS

N.B. When a reference is indicated by [20;.], it means the reference [.] of [20].

CONTENTS

CHAPTER IV : CHARACTERIZATION OF LOCALLY CONVEX PROPERTIES OF THE $C_P(X;E)$ SPACES

CHAPTER V : BOUNDED LINEAR FUNCTIONALS ON $C_P(X;E)$

CHAPTER I

SPACES OF VECTOR-VALUED CONTINUOUS FUNCTIONS

I.1. NOTATIONS

Unless explicitely stated otherwise,

$$X$$

is a Hausdorff completely regular space,

$$E$$

is a Hausdorff locally convex topological vector space and

$$C(X;E)$$

is the space of the continuous functions on X with values in E. In the case when X is equal to $\mathbb{R}$ (or $\mathbb{C}$), we simply write

$$C(X)$$

instead of $C(X;\mathbb{R})$ (or $C(X;\mathbb{C})$). As usual E' denotes the (topological) dual of E, E'_s denotes the weak*-dual of E, i.e. E' endowed with the $\sigma(E',E)$ topology, and E'_b denotes the strong dual of E, i.e. E' endowed with the $\beta(E',E)$ topology.

To simplify as much as possible the understanding of the symbols, we designate by letters such as

x, y the elements of X,
f, g the elements of C(X),
e, d the elements of E,
e', d' the elements of E',

φ, ψ the elements of $C(X;E)$.

Moreover

$$P$$

is the system of the continuous semi-norms on E and

$$Q$$

a system of semi-norms on E, finer than P.

<u>Remarks</u> I.1.1.

a) For every $f \in C(X)$, $e \in E$, $\varphi \in C(X;E)$, $p \in P$ and $e' \in E'$, we certainly have

$$fe,\ f\varphi \in C(X;E)$$

and

$$p(\varphi) = p\circ\varphi,\ \langle\varphi,e'\rangle = e'\circ\varphi \in C(X).$$

b) We always have

$$C(X;(E,Q)) \subset C(X;E).$$

Let us insist on the fact that, for $\varphi \in C(X;E)$, $q \in Q$ and $e' \in (E,Q)'$, the functions $q(\varphi)$ and $\langle\varphi,e'\rangle$ are not necessarily continuous on X.

<u>Remark</u> I.1.2. Given $\varphi \in C(X;E)$ and $p \in P$, we know that $\varphi \in C(X;E)$ [resp. $p \in C(E)$; $p(\varphi) \in C(X)$] has a unique continuous extension φ^{υ} [resp. p^{υ}; $p(\varphi)^{\upsilon}$] on υX [resp. υE; υX] with values in υE [resp. $\upsilon[0,+\infty[\ = [0,+\infty[$; $\upsilon[0,+\infty[\ = [0,+\infty[$). Therefore from the obvious equalities

$$p(\varphi)^{\upsilon}(x) = p(\varphi(x)) = p(\varphi^{\upsilon}(x)) = p^{\upsilon}(\varphi^{\upsilon}(x)), \quad \forall\ x \in X,$$

we get immediately

$$p(\varphi)^{\upsilon} = p^{\upsilon}(\varphi^{\upsilon}).$$

CONVENTION I.1.3. *We shall use the same symbol to designate a continuous function on a Hausdorff completely regular space with values into another one and its unique continuous extension in between the realcompactifications of these spaces.*

In particular, this convention implies that we give birth to the following equality

$$(p(\varphi))(x) = p(\varphi(x)),\ \forall p \in P,\ \forall \varphi \in C(X;E),\ \forall x \in \upsilon X.$$

It is however important to keep in mind that, for $x \in \upsilon X \setminus X$, $\varphi(x) \in \upsilon E$ may very well not belong to E although $p(\varphi(x))$ is a real number larger than or equal to zero.

I.2. The $C_P(X;E)$ spaces

Of course $C(X;E)$ is a linear space which can be endowed with many different locally convex topologies. In this paragraph we describe those topologies that we shall consider later on, i.e. the locally convex topologies on $C(X;E)$ of uniform convergence on subsets of υX.

PROPOSITION I.2.1.

a) *Given* $p \in P$ *and a subset* A *of* υX,

$$\sup_{x \in A} p(.(x))$$

is a semi-norm on $C(X;E)$ *if and only if* A *is a relatively compact subset of* υX.

b) *Given* $p_1, p_2 \in P$, *relatively compact subsets* B_1, B_2 *of* υX *and* $C > 0$, *one has*

$$\sup_{x \in B_1} p_1(\varphi(x)) \leq C \sup_{x \in B_2} p_2(\varphi(x)),\ \forall \varphi \in C(X;E),$$

if and only if $B_1 \subset \overline{B_2}^{\upsilon X}$ *and* $p_1 \leq Cp_2$ *hold*.

Proof. a) The condition is necessary. If A is not relatively compact in υX, it is not bounding (= "bornant" in [20]) and therefore there is an element f of $C(X)$ which is not bounded on A. Now let $e \in E$

be such that $p(e) = 1$, then fe belongs to $C(X;E)$ and satisfies

$$\sup_{x\in A} p(f(x)e) = \sup_{x\in A} |f(x)| = +\infty ,$$

which is sufficient.

The sufficiency of the condition comes directly from the fact that, for every $\varphi \in C(X;E)$, $p(\varphi)$ is a continuous function on X with values in $[0,+\infty[$ (by use of the convention I.1.3, $p(\varphi)$ is also a continuous function on υX with values in $[0,+\infty[$).

b) The condition is necessary. On one hand, let us prove the inclusion $B_1 \subset \overline{B_2}^{\upsilon X}$. If it is not the case, there are $x_o \in B_1$, $f \in C(X)$ and $e \in E$ such that $f(x_o) = 1$, $f(B_2) = \{0\}$ and $p_1(e) = 1$. This implies $fe \in C(X;E)$,

$$\sup_{x\in B_1} p_1(f(x)e) \geq 1$$

and

$$\sup_{x\in B_2} p_2(f(x)e) = 0$$

whence a contradiction. On the other hand, we have $p_1 \leq Cp_2$. If it is not the case, there is $e \in E$ such that $p_1(e) > Cp_2(e)$ and the consideration of $\chi_X e$ leads then directly to a contradiction.

The sufficiency of the condition is trivial.///

NOTATION I.2.2. For every $p \in P$ and every relatively compact subset B of υX, we designate by

$$p_B$$

the semi-norm

$$\sup_{x\in B} p(.(x))$$

on $C(X;E)$.

THEOREM I.2.3. *If* $\mathcal{P}$ *is a family of relatively compact subsets of* υX, *then*

$$\{p_B : p \in P, B \in \mathcal{P}\}$$

is a system of semi-norms on $C(X;E)$ *if and only if one may consider the space* $C_{\mathcal{P}}(X)$, i.e. (cf. [20], Proposition II.7.1.) if and only if the following two conditions are satisfied :

a) $\cup \{B : B \in \mathcal{P}\}$ is dense in υX,

b) for every $B_1, B_2 \in \mathcal{P}$, there is $B \in \mathcal{P}$ such that $B_1 \cup B_2 \subset \overline{B}^{\upsilon X}$.

Proof. The condition is necessary. On one hand, if part a) of the condition does not hold, there are $x_o \in \upsilon X$ and $f_o \in C(X)$ such that $f_o(x) = 1$ and $f_o(\cup \{B : B \in \mathcal{P}\}) = \{0\}$. As there are $e_o \in E$ and $p_o \in P$ such that $p_o(e_o) = 1$, $f_o e_o$ is a non zero element of $C(X;E)$ such that $p_B(f_o e_o) = 0$ for every $p \in P$ and every $B \in \mathcal{P}$, whence a contradiction. On the other hand, let B_1, B_2 be two elements of $\mathcal{P}$. For any $p \in P$, p_{B_1} and p_{B_2} are two elements of the system of semi-norms $\{p_B : p \in P, B \in \mathcal{P}\}$. Therefore there are $q \in P$ and $B \in \mathcal{P}$ such that

$$\sup \{p_{B_1}, p_{B_2}\} \leqslant Cq_B$$

hence such that

$$B_1 \subset \overline{B}^{\upsilon X} \quad \text{and} \quad B_2 \subset \overline{B}^{\upsilon X}$$

by proposition I.2.1, which is sufficient.

Of course the condition is sufficient.///

DEFINITION I.2.4. Unless explicitely stated, the notation

$$C_{\mathcal{P}}(X;E)$$

is used only if $\mathcal{P}$ is a family of relatively compact subsets of υX, defining a system of semi-norms on $C(X)$ [or, as we have just proved, on $C(X;E)$];

$$|\cdot|_{\mathcal{P}} \quad [\text{resp. } P_{\mathcal{P}}]$$

designates then the system of semi-norms

$$|\cdot|_{\mathcal{P}} = \{|\cdot|_B : B \in \mathcal{P}\} \quad [\text{resp. } P_{\mathcal{P}} = \{p_B : p \in P,\ B \in \mathcal{P}\}]$$

on C(X) [resp. C(X;E)].

We then set

$$C_{\mathcal{P}}(X;E) = [C(X;E), P_{\mathcal{P}}];$$

$C_{\mathcal{P}}(X;E)$ is therefore always a Hausdorff locally convex topological vector space.

EXAMPLES I.2.5.

a) <u>The space</u> C(X;E) <u>with the compact-open topology</u> is the space $C_{\mathcal{K}(X)}(X;E)$; the usual notation is

$$C_c(X;E).$$

[Let us recall that $\mathcal{K}(X)$ is the family of all the relatively compact subsets of X.]

b) <u>The space</u> C(X;E) <u>with the pointwise topology</u> is the space $C_{\mathcal{A}(X)}(X;E)$; the usual notation is

$$C_s(X;E).$$

[Let us recall that $\mathcal{A}(X)$ is the family of all the finite subsets of X.]

c) <u>The space</u> C(X;E) <u>with the bounding topology</u> is the space $C_{\mathcal{B}(X)}(X;E)$; the usual notation is

$$C_b(X;E).$$

[Let us recall that $\mathcal{B}(X)$ is the family of all the bounding subsets of X; a subset B of X is bounding if every continuous function on X is bounded on B. Of course a bounding subset of X is relatively compact in υX but may very well not be relatively compact in X.]

<u>Remark</u> I.2.6. In order to have uniform notations, we will systematically use the notation

$$C_{\mathcal{P}}(X) \quad [\text{resp. } C^b_{\mathcal{Q}}(X)]$$

instead of the notation $[C(X),\mathcal{P}]$ {resp. $[C^b(X),\mathcal{Q}]$} of [20].

I.3. THE $C_{Q,\mathcal{P}}(X;E)$ SPACES

The study of the $C_{\mathcal{P}}(X;E)$ spaces uses the introduction of the following spaces, which generalize them in some sense.

CONVENTION I.3.1. Unless explicitely stated, the notation

$$C_{Q,\mathcal{P}}(X;E)$$

is used only if the following conditions on $\mathcal{P}$ and Q are satisfied :

a) $\mathcal{P}$ is a family of bounding subsets of X such that one can consider the space $C_{\mathcal{P}}(X)$,

b) for every $B \in \mathcal{P}$, $\overline{B}^X$ belongs to $\mathcal{P}$,

c) Q (as stated in paragraph I.1) is a system of semi-norms on E, finer than P,

d) for every $q \in Q$, $B \in \mathcal{P}$ and $\varphi \in C(X;E)$, one has

$$\sup_{x \in B} q(\varphi(x)) < \infty .$$

<u>Remark</u> I.3.2. Given $C_{Q,\mathcal{P}}(X;E)$,

$$Q_{\mathcal{P}} = \{\sup_{x \in B} q(.(x)) : q \in Q, B \in \mathcal{P}\}$$

is of course a system of semi-norms on C(X;E), finer than the one of $C_{\mathcal{P}}(X;E)$.

DEFINITION I.3.3. The notation

$$C_{Q,\mathcal{P}}(X;E)$$

designates then the Hausdorff locally convex topological vector space obtained by endowing the linear space C(X;E) with the system of semi-norms $Q_{\mathcal{P}}$.

Notations such as

$$C_{Q,c}(X;E) \quad \text{and} \quad C_{Q,s}(X;E)$$

are clear [let us remark however that the first one may be used only if for every $q \in Q$ and $\varphi \in C(X;E)$, the function $q(\varphi)$ on X is bounded on every compact subset of X; the second is always defined].

EXAMPLES I.3.4.

a) If Q is equivalent to P, the spaces $C_{Q,P}(X;E)$ and $C_P(X;E)$ are of course topologically isomorphic (and nothing is gained). [Let us insist on the fact that in the notation $C_P(X;E)$, P is a family of relatively compact subsets of υX while in the notation $C_{Q,P}(X;E)$, P is a family of bounding subsets of X.]

b) The important examples of $C_{Q,P}(X;E)$ spaces are given by the following two cases :

i) $P = A(X)$, Q general : we get the space $C_{Q,s}(X;E)$

ii) $P = K(X)$ or $B(X)$, $Q = P_b$ [let us recall that P_b is the system of semi-norms of the bornological space associated to E] : we get the spaces $C_{P_b,c}(X;E)$ or $C_{P_b,b}(X;E)$.

I.4. SPECIAL SUBSPACES OF $C_P(X;E)$

DEFINITION I.4.1. For every $x \in X, f \in C(X)$, $e \in E$ and $e' \in E'$, we introduce the following applications

$$\delta_x \; : \; C(X;E) \to E$$

$$\delta_f \; : \; E \to C(X;E)$$

$$\delta_e \; : \; C(X) \to C(X;E)$$

$$\delta_{e'} \; : \; C(X;E) \to C(X)$$

defined respectively by

$$\delta_x \varphi = \varphi(x), \quad \forall \varphi \in C(X;E),$$

$$\delta_f e = fe \quad , \quad \forall e \in E,$$

$$\delta_e f = fe \quad , \qquad \forall f \in C(X),$$

$$\delta_{e'} \varphi = e' \circ \varphi , \qquad \forall \varphi \in C(X;E).$$

The following properties of these maps are immediate.

PROPOSITION I.4.2.

a) For every $x \in X$, δ_x is a linear map from $C(X;E)$ into E. If moreover x belongs to $\cup \{\overline{B}^{\cup X^x} : B \in P\}$, then δ_x is a continuous linear map from $C_P(X;E)$ into E [resp. from $C_{Q,P}(X;E)$ into (E,Q)].

b) For every non zero $f \in C(X)$, δ_f is a topological isomorphism in between E and the linear subspace $\delta_f E$ of $C_P(X;E)$ [resp. in between (E,Q) and the linear subspace $\delta_f E$ of $C_{Q,P}(X;E)$].

c) For every non zero $e \in E$, δ_e is a topological isomorphism in between $C_P(X)$ and the linear subspace $\delta_e C(X)$ of $C_P(X;E)$ [resp. of $C_{Q,P}(X;E)$].

d) For every $e' \in E'$, $\delta_{e'}$ is a continuous linear map from $C_P(X;E)$ into $C_P(X)$ and from $C_{Q,P}(X;E)$ into $C_P(X)$.///

PROPOSITION I.4.3.

a) If $x \in X$ and $f \in C(X)$ are such that $f(x) = 1$, the following equalities hold

$$\begin{cases} \delta_f \circ (\delta_x |_{\delta_f E}) = \mathrm{Id}_{\delta_f E} \\ \delta_x \circ \delta_f = \mathrm{Id}_E . \end{cases}$$

Therefore if moreover x belongs to $\cup \{\overline{B}^{\cup X} : B \in P\}$, then $\delta_f \circ \delta_x$ is a continuous linear projection from $C_P(X;E)$ onto $\delta_f E$ and from $C_{Q,P}(X;E)$ onto $\delta_f E$. In particular, $\delta_f E$ is then a topologically complemented subspace of $C_P(X;E)$ and of $C_{Q,P}(X;E)$.

b) If $e \in E$ and $e' \in E'$ are such that $\langle e,e' \rangle = 1$, the following equalities hold

$$\begin{cases} \delta_e \circ (\delta_{e'} |_{\delta_e C(X)}) = \mathrm{Id}_{\delta_e C(X)} \\ \delta_{e'} \circ \delta_e = \mathrm{Id}_{C(X)} . \end{cases}$$

Therefore $\delta_e \circ \delta_{e'}$ is a continuous linear projection from $C_P(X;E)$

onto $\delta_e C(X)$. *In particular, for every non zero* $e \in E$, $\delta_e C(X)$ *is a topologically complemented subspace of* $C_{\mathcal{P}}(X;E)$.///

We then get the following result as a direct consequence of these previous two propositions.

THEOREM I.4.4.

a) *If some point of* X *belongs to* $\cup \{\overline{B}^{\upsilon X} : B \in \mathcal{P}\}$, *then* E *is topologically isomorphic to a topologically complemented subspace of* $C_{\mathcal{P}}(X;E)$.

The space (E,Q) *is always topologically isomorphic to a topologically complemented subspace of* $C_{Q,\mathcal{P}}(X;E)$.

b) *The space* $C_{\mathcal{P}}(X)$ *is always topologically isomorphic to a topologically complemented subspace of* $C_{\mathcal{P}}(X;E)$.///

I.5. $C(X)\otimes E$ IS DENSE IN $C_c(X;E)$

PROPOSITION I.5.1. *For every disjoint compact subset* K *and closed subset* F *of* X, *there is* $f \in C(X)$ *with values in* $[0,1]$ *which is identically* 1 *on a neighborhood of* K *and* 0 *on a neighborhood of* F.

Proof. As X is Hausdorff and completely regular, there is $g \in C(X)$ with values in $[0,1]$ which is identically 1 on K and 0 on F. Then of course

$$V_1 = \{x \in X : g(x) \geq 3/4\} \quad \text{and} \quad V_2 = \{x \in X : g(x) \leq 1/4\}$$

are neighborhoods of K and F respectively. Now it is trivial matter to check that one can take $f = h \circ g$ if h is a continuous function on $[0,1]$, with values in $[0,1]$ which is identically 1 on $[3/4,1]$ and 0 on $[0,1/4]$.///

PROPOSITION I.5.2. *If* $\{\Omega_j : j \leq J\}$ *is a finite open covering of a compact subset* K *of* X, *there are continuous functions* f_j $(j \leq J)$ *on* X *with values in* $[0,1]$, *such that* $\operatorname{supp} f_j \subset \Omega_j$ *for every* $j \leq J$ *and that* $\Sigma_{j=1}^{J} f_j$ *is identically* 1 *on a neighborhood of* K *and bounded by* 1 *on* X.

<u>Proof</u>. Of course

$$K_1 = K \setminus \bigcup_{j=2}^{J} \Omega_j$$

is a compact subset of Ω_1. Therefore by the preceding result, there is $g_1 \in C(X)$ with values in $[0,1]$ which is identically 1 on an open neighborhood V_1 of K_1 and 0 on a neighborhood of $X \setminus \Omega_1$. In particular, the support of g_1 is contained in Ω_1.

By induction on $l = 2, \ldots, J-1$, we see that

$$K_l = (K \setminus \bigcup_{j=l+1}^{J} \Omega_j) \setminus (V_1 \cup \ldots \cup V_{l-1})$$

is a compact subset of Ω_l and get the existence of $g_l \in C(X)$ with values in $[0,1]$ which is identically 1 on an open neighborhood V_l of K_l and with support contained in Ω_l.

Finally

$$K_J = K \setminus (V_1 \cup \ldots \cup V_{J-1})$$

is a compact subset of Ω_J. Therefore there is $g_J \in C(X)$ with values in $[0,1]$ which is identically 1 on K_J and with support contained in Ω_J.

As $\Sigma_{j=1}^{J} g_j$ is continuous on X and larger than or equal to 1 at every point of K,

$$V = \{x \in X : \sum_{j=1}^{J} g_j(x) > \tfrac{1}{2}\}$$

is an open neighborhood of K. Therefore there is $g \in C(X)$ with values in $[0,1]$ which is identically 1 on a neighborhood of K and with support contained in V.

It is then easy to check that one can take the functions f_j $(j \leq J)$ defined by

$$f_j(x) = \left\{\begin{array}{ll} g(x)g_j(x) \;/\; [\sum_{k=1}^{J} g_k(x)] & \text{if } x \in V \\ 0 & \text{if } x \in X \setminus V \end{array}\right\}. ///$$

THEOREM I.5.3. *For every* $\varphi \in C(X;E)$, *every compact subset* K *of* X, *every* $p \in P$ *and every* $\varepsilon > 0$, *there is a finite continuous partition of unity* $\{f_1, \ldots, f_J\}$ *on* X *and a finite subset* $\{x_1, \ldots, x_J\}$ *of* K *such that*

$$p_K(\varphi - \sum_{j=1}^{J} f_j\varphi(x_j)) \leq \varepsilon.$$

In particular, $C(X)\otimes E$ *is always a dense linear subspace of* $C_c(X;E)$.

Proof. It is direct to get a finite open covering $\{\Omega_j : j \leq J-1\}$ of K such that, for every $j \leq J-1$, one has

$$x, y \in \Omega_j\cap K \quad \Rightarrow \quad p(\varphi(x) - \varphi(y)) \leq \varepsilon.$$

By use of the preceding result, there are then continuous functions f_j $(j \leq J-1)$ on X with values in $[0,1]$, such that $\operatorname{supp} f_j \subset \Omega_j$ for every $j \leq J-1$ and that $\Sigma_{j=1}^{J-1} f_j$ is identically 1 on a neighborhood of K and bounded by 1 on X. For every $j \leq J-1$, let us choose a point x_j of $K \cap \Omega_j$.

Moreover let us set $f_J = \chi_X - \Sigma_{j=1}^{J-1} f_j$ and choose any point x_J of K. Then f_J belongs to $C(X)$ and is identically 0 on K.

Therefore we get

$$p_K(\varphi - \sum_{j=1}^{J} f_j\varphi(x_j)) \leq \sup_{x\in K} \sum_{j=1}^{J-1} f_j(x)p(\varphi(x) - \varphi(x_j))$$

$$\leq \sup_{x\in K} \sum_{j=1}^{J-1} f_j(x)\varepsilon = \varepsilon.$$

Hence the conclusion.///

I.6. THE SETS C(X;D) [17]

NOTATIONS I.6.1. For every subset D of E, we set

$$C_E(X;D) = \{\varphi \in C(X;E) : \varphi(X) \subset D\},$$

the more convenient notation

$$C(X;D)$$

will be used whenever no confusion is possible.

If D is a subset of E,

$$P_E(X;D)$$

designates the set of the linear combinations of the type $\Sigma_{j=1}^{J} f_j e_j$ where J is an integer, where the e_j belong to D and where $\{f_1, \ldots, f_J\}$ is a finite continuous partition of unity on X. When no confusion is possible, we simply write

$$P(X;D).$$

For a compact space K, $p \in P$ and $r > 0$, let us remark that the semi-ball

$$\{\varphi \in C_c(K;E) : p_K(\varphi) \leq r\}$$

can be written $C(K;b_p(r))$.

PROPOSITION I.6.2. <u>For every absolutely convex subset</u> D <u>of</u> E, <u>we have</u>

a) $P(X;D) \subset C(X;D)$,

b) $\overline{C(X;D)} = \overline{P(X;D)}$,

c) $\overline{P(X;D)} = \overline{P(X;\overline{D})}$,

d) $C(X;\overline{D}) = \overline{C(X;\overline{D})} = \overline{C(X;D)}$,

<u>where all the closures are taken in</u> $C_c(X;E)$.

<u>Proof</u>. a) is trivial.

b) By use of a), it is enough to prove that $C(X;D)$ is contained in $\overline{P(X;D)}$, which is directly implied by theorem I.5.3.

c) Of course it is enough to prove that $\overline{P(X;\overline{D})}$ is contained in $\overline{P(X;D)}$. Let φ belong to $\overline{P(X;\overline{D})}$. For every compact subset K of X, $p \in P$ and $r > 0$, by theorem I.5.3, there are $J \in \mathbb{N}$, $e_j \in \overline{D}$ $(j \leq J)$

and a finite continuous partition of unity $\{f_1, \ldots, f_J\}$ on X such that

$$p_K(\varphi - \sum_{j=1}^{J} f_j e_j) \leq \frac{r}{2}.$$

For every $j \leq J$, there is then $d_j \in D$ such that $p(e_j - d_j) \leq r/2$. Therefore we get

$$p_K(\varphi - \sum_{j=1}^{J} f_j d_j)$$

$$\leq p_K(\varphi - \sum_{j=1}^{J} f_j e_j) + \sup_{x \in K} \sum_{j=1}^{J} f_j(x) p(e_j - d_j) \leq \varepsilon.$$

Hence the conclusion.

d) To get the first equality it is enough to prove that $C(X;\bar{D})$ is a closed subset of $C_c(X;E)$. This is a direct consequence of part a) of proposition I.4.2 : for every $x \in X$, δ_x is a continuous linear map from $C_c(X;E)$ into E and we have

$$C(X;\bar{D}) = \bigcap_{x \in X} (\delta_x)^{-1}(\bar{D}).$$

The second equality is a direct consequence of b) and c) :

$$\overline{C(X;\bar{D})} = \overline{P(X;\bar{D})} = \overline{P(X;D)} = \overline{C(X;D)}.///$$

PROPOSITION I.6.3. <u>For every</u> $J \in \mathbb{N}$ <u>and every absolutely convex subsets</u> $D_1, \ldots, D_J$ <u>of</u> E, <u>we have</u>

a) $P(X;\Sigma_{j=1}^{J} D_j) = \sum_{j=1}^{J} P(X;D_j)$,

b) $C(X;\Sigma_{j=1}^{J} D_j) \supset \sum_{j=1}^{J} C(X;D_j)$,

c) $\overline{C(X;\Sigma_{j=1}^{J} D_j)} = \overline{\Sigma_{j=1}^{J} C(X;D_j)}$,

<u>where all the closures are taken in</u> $C_c(X;E)$.

<u>Proof</u>. a) Of course it is sufficient to prove this equality for $J = 2$. The inclusion

$$P(X;D_1 + D_2) \subset P(X;D_1) + P(X;D_2)$$

is immediate. To get the other inclusion, let us remark that if $\{f_l : l \leq L\}$ and $\{g_k : k \leq K\}$ are finite continuous partitions of unity on X, then

$$\{f_l g_k : f_l g_k \neq 0,\ l \leq L,\ k \leq K\}$$

is a finite continuous partition of unity on X such that

$$\sum_{l=1}^{L} f_l e_{1,l} + \sum_{k=1}^{K} g_k e_{2,k} = \sum_{l=1}^{L} \sum_{k=1}^{K} f_l g_k (e_{1,l} + e_{2,k})$$

for every $e_{1,1}, \ldots, e_{1,L} \in D_1$ and $e_{2,1}, \ldots, e_{2,K} \in D_2$.

b) is immediate.

c) It is enough to check that we successively have

$$\overline{C(X;\Sigma_{j=1}^{J} D_j)} = \overline{C(X;\Sigma_{j=1}^{J} D_j)} = \overline{P(X;\Sigma_{j=1}^{J} D_j)} = \overline{\Sigma_{j=1}^{J} P(X;D_j)}$$

$$\subset \overline{\Sigma_{j=1}^{J} C(X;D_j)} \subset \overline{C(X;\Sigma_{j=1}^{J} D_j)} = \overline{C(X;\Sigma_{j=1}^{J} D_j)}.///$$

PROPOSITION I.6.4. <u>For any increasing sequence</u> $(D_m)_{m\in\mathbb{N}}$ <u>of absolutely convex subsets of</u> E, <u>we have</u>

a) $P(X; \bigcup_{m=1}^{\infty} D_m) = \bigcup_{m=1}^{\infty} P(X;D_m)$,

b) $\bigcup_{m=1}^{\infty} C(X;D_m) \subset C(X; \bigcup_{m=1}^{\infty} D_m)$,

c) $C(X; \overline{\cup_{m=1}^{\infty} D_m)} = \overline{\cup_{m=1}^{\infty} C(X;D_m)}$,

<u>where all the closures are taken in</u> $C_c(X;E)$.

<u>Proof</u>. a) and b) are immediate.

c) It is enough to check that we successively have

$$C(X; \overline{\cup_{m=1}^{\infty} D_m)} = \overline{C(X; \cup_{m=1}^{\infty} D_m)} = \overline{P(X; \cup_{m=1}^{\infty} D_m)}$$

$$= \overline{\cup_{m=1}^{\infty} P(X;D_m)} \subset \overline{\cup_{m=1}^{\infty} C(X;D_m)} \subset \overline{C(X; \cup_{m=1}^{\infty} D_m)}.///$$

Here is a fine property of these sets $C(X;D)$.

PROPOSITION I.6.5. Let K be a compact space. If E has a fundamental sequence of bounded subsets then, for every bornivorous barrel T of $C_c(K;E)$, there is a bornivorous barrel T_1 of E such that $C(K;T_1) \subset T$.

Proof. Of course E has then a fundamental sequence B_m of bounded subsets which are absolutely convex and increasing. As the sets $C(K;B_m)$ $(m \in \mathbb{N})$ are all bounded in $C_c(K;E)$, there is a sequence $r_m > 0$ such that

$$C(K;B_m) \subset \frac{1}{r_m} T, \quad \forall\, m \in \mathbb{N},$$

hence such that

$$C(K;2^{-m} r_m B_m) \subset 2^{-m} T, \quad \forall\, m \in \mathbb{N}.$$

Then we get successively

$$C\Big(K; \overline{\bigcup_{m=1}^{\infty} \Big(\sum_{k=1}^{m} 2^{-k} r_k B_k\Big)}\Big) = \overline{\bigcup_{m=1}^{\infty} C\Big(K; \sum_{k=1}^{m} 2^{-k} r_k B_k\Big)}$$

$$\subset \overline{\bigcup_{m=1}^{\infty} C\Big(K; \overline{\textstyle\sum_{k=1}^{m} 2^{-k} r_k B_k}\Big)}$$

$$\subset \overline{\bigcup_{m=1}^{\infty} \overline{\sum_{k=1}^{m} C(K;2^{-k} r_k B_k)}}$$

$$\subset \overline{\bigcup_{m=1}^{\infty} \sum_{k=1}^{m} 2^{-k} T} \subset T$$

and to conclude, it is enough to check that

$$T_1 = \overline{\bigcup_{m=1}^{\infty} \Big(\sum_{k=1}^{m} 2^{-k} r_k B_k\Big)}$$

is a bornivorous barrel of E.///

I.7. THE MUJICA THEOREM ABOUT $C_c(X;\text{Ind } E_m)$ [19]

CONVENTION I.7.1. Unless explicitely stated, in this paragraph

$$(E_m)_{m\in\mathbb{N}}$$

is a sequence of Hausdorff locally convex topological vector spaces such that $E_m \subset E_{m+1}$ for every $m \in \mathbb{N}$. Moreover for every $m \in \mathbb{N}$,

$$P_m$$

is the system of all the continuous semi-norms on E_m and we suppose that P_{m+1} is coarser than P_m on E_m. Finally

$$\text{Ind } E_m$$

is the inductive limit of these spaces E_m and _we suppose that_ Ind E_m _is a Hausdorff space_.

It is known that the locally convex topology of Ind E_m can be described by the system of the semi-norms $p_{\pi,\gamma}$ defined in the following way : for every sequence $\pi = (p_m)_{m\in\mathbb{N}}$ where for every m, p_m belongs to P_m and for every sequence $\gamma = (r_m)_{m\in\mathbb{N}}$ of strictly positive numbers, $p_{\pi,\gamma}$ is defined on Ind E_m by

$$p_{\pi,\gamma}(e) = \inf_{\substack{e=\Sigma_{(j)} e_j \\ e_j\in E_j}} \Sigma_{(j)} r_j p_j(e_j)$$

where $\Sigma_{(j)}$ indicates that the sum is finite.

THE MUJICA THEOREM I.7.2. _If every countable union of relatively compact subsets of_ X _is relatively compact in_ X, _then_ Ind $C_c(X;E_m)$ _is a dense topological subspace of_ $C_c(X;\text{Ind } E_m)$.

Proof. Let us establish first that it makes sense to consider the inductive limit of the spaces $C_c(X;E_m)$ and that this inductive limit is Hausdorff. On one hand, for every $m \in \mathbb{N}$, E_m is contained in E_{m+1} and has a finer topology than the one induced by E_{m+1}. Therefore for every $m \in \mathbb{N}$, $C_c(X;E_m)$ is contained in $C_c(X;E_{m+1})$ and has a finer topology than the one induced by $C_c(X;E_{m+1})$. On the other hand, let φ

be an element of $\cup_{m=1}^{\infty} C_c(X;E_m)$ which is annihilated by all the continuous semi-norms of Ind $C_c(X;E_m)$. For every $x \in X$, $\{x\}$ is a compact subset of X. Therefore for every sequences $\pi = (p_m)_{m\in\mathbb{N}}$ and $\gamma = (r_m)_{m\in\mathbb{N}}$, we have

$$\inf_{\substack{\varphi=\Sigma_{(j)}\varphi_j \\ \varphi_j\in C_c(X;E_j)}} \sum_{(j)} r_j p_{j,\{x\}}(\varphi_j) = \inf_{\substack{\varphi(x)=\Sigma_{(j)}e_j \\ e_j\in E_j}} \sum_{(j)} r_j p_j(e_j) = 0,$$

i.e. $\varphi(x) = 0$ since Ind E_m is Hausdorff.

At this point, the inclusion

$$\text{Ind } C_c(X;E_m) \subset C_c(X;\text{Ind } E_m)$$

is immediate.

Now let us prove that Ind $C_c(X;E_m)$ has the topology induced by $C_c(X;\text{Ind } E_m)$. In fact the identity map

$$I : \text{Ind } C_c(X;E_m) \to C_c(X;\text{Ind } E_m)$$

is

a) of course linear and injective.

b) continuous as one can check by direct calculation.

c) relatively open. This we prove in the next paragraph by showing that, for every closed semi-ball

$$b = \{\varphi \in \text{Ind } C_c(X;E_m) : \inf_{\substack{\varphi=\Sigma_{(j)}\varphi_j \\ \varphi_j\in C_c(X;E_j)}} \sum_{(j)} r_j \sup_{x\in K_j} p_j(\varphi_j(x)) \leq 1\}$$

of Ind $C_c(X;E_m)$, we have

$$b' \cap \bigcup_{m=1}^{\infty} C_c(X;E_m) \subset b \qquad (\star)$$

where b' is the following semi-ball of $C_c(X;\text{Ind } E_m)$

$$b' = \{\varphi \in C_c(X;\text{Ind } E_m) : \sup_{x\in K} \inf_{\substack{\varphi(x)=\Sigma_{(j)}e_j \\ e_j\in E_j}} \sum_{(j)} 2^j r_j p_j(e_j) \leq \tfrac{1}{2}\},$$

where K is a compact subset of X containing $\cup_{j=1}^{\infty} K_j$.

To prove the inclusion (★), it is sufficient to establish that for every $\varphi \in b' \cap C_c(X;E_m)$, every continuous semi-norm $p_{\pi',\gamma'}$ on Ind $C_c(X;E_m)$ and every $r > 0$, there is $\psi \in b$ such that $p_{\pi',\gamma'}(\varphi - \psi) \leq r$ because b is closed. It is direct matter to find a finite number of points $x_1, \ldots, x_L$ of K such that the open subsets of X

$$\Omega_l = \{x \in X : r'_m p'_m(\varphi(x) - \varphi(x_l)) < r\} \quad (l \leq L)$$

constitute a finite open cover of K. By proposition I.5.2, we get functions $f_l \in C(X)$ $(l \leq L)$ with values in [0,1] such that supp $f_l \subset \Omega_l$ for every $l \leq L$ and that $\Sigma_{l=1}^{L} f_l$ is identically 1 on a neighborhood of K and bounded by 1 on X. By the definition of b', there is then an integer J larger than or equal to m and elements $e_{l,j}$ of E_j $(l \leq L\ ;\ j \leq J)$ such that

$$\varphi(x_l) = \sum_{j=1}^{J} e_{l,j}, \quad \forall\, l \leq L,$$

and

$$\sum_{j=1}^{J} 2^j r_j p_j(e_{l,j}) \leq 1, \quad \forall\, l \leq L.$$

If we set

$$\psi'_j = \sum_{l=1}^{L} f_l e_{l,j}, \quad \forall\, j \leq J,$$

and

$$\psi' = \sum_{j=1}^{J} \psi'_j = \sum_{l=1}^{L} f_l \sum_{j=1}^{J} e_{l,j} = \sum_{l=1}^{L} f_l \varphi(x_l),$$

we get $\psi'_j \in C_c(X;E_j)$ for every $j \leq J$, $\psi' \in C_c(X;E_m)$ and

$$r'_m p'_m(\varphi(x) - \psi'(x)) < r, \quad \forall\, x \in K.$$

Therefore there is an open neighborhood V of K such that

$$r'_m p'_m(\varphi(x) - \psi'(x)) < r, \quad \forall\, x \in V.$$

Now we can choose $f \in C(X)$ with values in $[0,1]$ and support contained in V which is identically 1 on K. Finally

$$\psi = f\psi' + (1 - f)\varphi$$

a) belongs to b since we have

$$\sum_{j=1}^{J} r_j \sup_{x\in K_j} p_j(f(x)\psi'_j(x)) + r_{J+1} \sup_{x\in K_{J+1}} p_{J+1}((1 - f(x))\varphi(x))$$

$$\leq \sum_{j=1}^{J} r_j \sup_{l\leq L} p_j(e_{l,j}) + 0 \leq \sum_{j=1}^{J} 2^{-j} \leq 1.$$

b) we have $p_{\pi',\gamma'}(\varphi - \psi) \leq r$ since we even have

$$r'_m p'_m(\varphi(x) - \psi(x)) = r'_m p'_m(f(x)(\varphi(x) - \psi'(x))) < r$$

for every $x \in X$.

Finally the density of Ind $C_c(X;E_m)$ into $C_c(X;\text{Ind } E_m)$ is a direct consequence of theorem I.5.3 since we certainly have

$$C(X)\otimes E \subset \text{Ind } C_c(X;E_m).///$$

COROLLARY I.7.3. _If_ K _is a Hausdorff compact space and if_ Ind E_m _is regularly compact_ (i.e. every compact subset of Ind E_m is contained and compact in one of the spaces E_m), _then the equality_

$$C_c(K;\text{Ind } E_m) = \text{Ind } C_c(K;E_m)$$

holds algebraically and topologically.///

I.8. THE $CRC_Q(X;E)$ SPACES

A priori the extension of the space $C^b(X)$ to the case of the vector-valued functions may be done in different ways. One may for instance consider the linear subspace of $C(X;E)$ of the elements $\varphi \in C(X;E)$ such that $\varphi(X)$ is relatively compact [resp. relatively countably compact; precompact; bounded; ...] in E. In this study the first case is very interesting and we shall limit ourselves to it.

DEFINITION I.8.1. The notation

$$CRC(X;E)$$

designates the set of the elements φ of $C(X;E)$ such that $\varphi(X)$ is relatively compact in E.

Of course $CRC(X;E)$ is a linear subspace of $C(X;E)$ and we may endow it with the locally convex topology induced by any of the spaces $C_P(X;E)$. The following remark will allow us to do better.

Remark I.8.2. Every $\varphi \in CRC(X;E)$ has a unique continuous extension φ^β from βX into βE. As $\varphi(X)$ is a relatively compact subset of E, $\varphi^\beta(\beta X)$ is in fact the compact subset $\overline{\varphi(X)}^E$ of E. In particular every $\varphi \in CRC(X;E)$ has a unique continuous extension from βX into E that (to simplify the notations as much as possible) we shall denote by φ too.

PROPOSITION I.8.3.

a) _For every_ $p \in P$ _and every subset_ A _of_ βX,

$$\sup_{x \in A} p(\varphi(x))$$

is a semi-norm on $CRC(X;E)$.

b) _Given_ $p_1, p_2 \in P$, _subsets_ A_1, A_2 _of_ βX _and_ $C > 0$, _one has_

$$\sup_{x \in A_1} p_1(\varphi(x)) \leqslant C \sup_{x \in A_2} p_2(\varphi(x)), \quad \forall \varphi \in CRC(X;E),$$

if and only if $A_1 \subset \overline{A_2}^{\beta X}$ _and_ $p_1 \leqslant Cp_2$ _hold_.

Proof. a) is immediate.

b) can be established by use of a similar argument as the one of part b) of proposition I.2.1.///

NOTATION I.8.4. For every $p \in P$ and $A \subset \beta X$, we denote

$$p_A$$

the semi-norm

$$p_A(.) = \sup_{x \in A} p(.(x))$$

on $CRC(X;E)$.

THEOREM I.8.5. <u>If</u> Q <u>is a family of subsets of</u> βX,

$$\{p_A : p \in P, A \in Q\}$$

<u>is a system of semi-norms on</u> $CRC(X;E)$ <u>if and only if one may consider the space</u> $C^b_Q(X)$, i.e. (cf. [20], Proposition II.7.4) if and only if the following two conditions are satisfied :

a) $\cup \{A : A \in Q\}$ is dense in βX,

b) for every $A_1, A_2 \in Q$, there is $A \in Q$ such that $A_1 \cup A_2 \subset \overline{A}^{\beta X}$.

<u>Proof</u>. The argument used to establish the theorem I.2.3 can be adapted easily to get this result.///

DEFINITION I.8.6. Unless explicitely stated, the notation

$$CRC_Q(X;E)$$

is used only if Q is a family of subsets of βX, defining a system of semi-norms on $C^b(X)$ [or, as we have just proved, on $CRC(X;E)$] ;

$$|.|_Q \quad [\text{resp. } P_Q]$$

designates then the system of semi-norms

$$|.|_Q = \{|.|_A : A \in Q\} \quad [\text{resp. } P_Q = \{p_A : p \in P, A \in Q\}]$$

on $C^b(X)$ [resp. $CRC(X;E)$].

We then set

$$CRC_Q(X;E) = [CRC(X;E), P_P] ;$$

$CRC_{Q}(X;E)$ is therefore always a Hausdorff locally convex topological vector space.

EXAMPLES I.8.7.

a) The space CRC(X;E) with the uniform convergence is the space $CRC_{p(X)}(X;E)$. The usual notation is

$$CRC_{u}(X;E).$$

[$p(X)$ is the family of all the subsets of X.] It is equivalent to consider the space $CRC_{p(\beta X)}(X;E)$.

b) As CRC(X;E) is a linear subspace of $C_{p}(X;E)$, one can of course introduce the space CRC(X;E) with the compact-open [resp. pointwise; bounding] topology. It is denoted by

$$CRC_{c}(X;E) \quad [\text{resp. } CRC_{s}(X;E) \; ; \; CRC_{b}(X;E)] .$$

In fact the spaces that we just introduced are not fundamentally new. The following result shows this; it will play an important role later on.

THEOREM I.8.8. The map β which, to every $\varphi \in CRC(X;E)$, associates the function

$$\beta\varphi = \varphi^{\beta} : \beta X \to E$$

is in fact an isomorphism in between

a) the linear spaces CRC(X;E) and $C(\beta X;E)$,

b) the Hausdorff locally convex topological vector spaces $CRC_{Q}(X;E)$ and $C_{Q}(\beta X;E)$, hence in particular in between the spaces $CRC_{u}(X;E)$ and $C_{c}(\beta X;E)$.

Proof. a) As $\varphi(X)$ is a relatively compact subset of E for every $\varphi \in CRC(X;E)$, it is enough to note that $\varphi^{\beta}(\beta X)$ is in fact the compact subset $\overline{\varphi(X)}^{E}$ of E.

b) is then immediate.///

CHAPTER II

ABSOLUTELY CONVEX SUBSETS OF C(X;E)

II.1. THE FUNDAMENTAL PROPERTY [20;35] & [21]

DEFINITION II.1.1. Let D be an absolutely convex subset of C(X;E). Then a hold of D is a compact subset K of βX such that $\varphi \in C(X;E)$ belongs to D if its unique continuous extension φ^{β} of βX into βE is identically 0 on a neighborhood of K.

Of course βX itself is a hold of every absolutely convex and non void subset of C(X;E).

[In fact this notion of hold could be introduced for general subsets of C(X;E). Then βX would be a hold of every subset of C(X;E) containing 0 and the following result would hold (with unchanged proof) for subsets of C(X;E) containing 0 as well as half the sum of any two of its elements. However this more general result would not provide more information later on.]

THE FUNDAMENTAL PROPERTY II.1.2. Every absolutely convex and non void subset of C(X;E) has a smallest hold.

In other words, every intersection of holds of an absolutely convex and non void subset D of C(X;E) is again a hold of D.

Proof. Let us establish first that the intersection of two holds K_1, K_2 of D is also a hold of D. Let φ be an element of C(X;E) such that φ^{β} is identically 0 on an open neighborhood V of $K_1 \cap K_2$ in βX. If V is a neighborhood of K_2, we already have $\varphi \in D$. If V is not a neighborhood of K_2, $K_2 \setminus V$ is a compact subset of βX disjoint from K_1. Therefore there is $f \in C(X)$ equal identically to 1 on a neighborhood V_1 of K_1 and to 0 on a neighborhood V_2 of $K_2 \setminus V$. Then on one hand $(2f\varphi)^{\beta}$ is identically 0 on $V \cup V_2$ hence belongs to D since $V \cup V_2$ is a neighborhood of K_2 and on the other hand $[2(1-f)\varphi]^\beta$ is identically 0 on V_1 hence belongs to D. Therefore

$$\varphi = \frac{1}{2}[2f\varphi + 2(1-f)\varphi]$$

belongs to D. Hence the conclusion.

From this we get at once that any finite intersection of holds of D is a hold of D.

Now let us prove that any intersection K of holds K_α ($\alpha \in A$) of D is a hold of D. Let φ be an element of C(X;E) such that φ^β is identically 0 on an open neighborhood V of K. Then there is a finite subset $\{\alpha_1, \ldots, \alpha_J\}$ of A such that the hold $K_{\alpha_1} \cap \ldots \cap K_{\alpha_J}$ of D is contained in V. Then φ belongs to D. Hence the conclusion.///

DEFINITION II.1.3. The _support_ of an absolutely convex and non void subset D of C(X;E) is the smallest hold of D; it is denoted by

$$K(D).$$

Here is an essential complement to the fundamental property. It provides a way to know if an absolutely convex subset of $C_P(X;E)$ is a neighborhood of 0.

THEOREM II.1.4. _Let_ D _be an absolutely convex subset of_ C(X;E). _If there are_ $p \in P$ _and_ $r > 0$ _such that_

$$D \supset \{\varphi \in C(X;E) : \sup_{x \in X} p(\varphi(x)) \leq r\}$$

then

a) _there is a smallest compact subset_ K _of_ βX _such that_ $\varphi \in C(X;E)$ _belongs to_ D _if_ φ^β _is identically_ 0 _on_ K. _Moreover this smallest compact subset coincides with the support_ K(D) _of_ D,

b) _one has_

$$\{\varphi \in C(X;E) : \sup_{x \in K(D)} p(\varphi)^\beta(x) < r\} \subset D.$$

Proof. a) Of course it is enough to prove that under the hypothesis a compact subset K of βX is a hold of D if and only if $\varphi \in C(X;E)$ belongs to D whenever φ^β is identically 0 on K.

The condition is necessary. Let K be a hold of D and let $\varphi \in C(X;E)$ be such that $\varphi^{\beta}(K) = \{0\}$. The map $\Theta : E \to E$ defined by

$$\Theta e = \begin{cases} e & \text{if} \quad p(e) \leqslant \frac{r}{2} \\ \dfrac{re}{2p(e)} & \text{if} \quad p(e) \geqslant \frac{r}{2} \end{cases}$$

is certainly continuous. It is then direct matter to verify that $2\ \Theta\circ\varphi$ belongs to D and that $2(\varphi - \Theta\circ\varphi)$ belongs to $C(X;E)$ and is such that $[2(\varphi - \Theta\circ\varphi)]^{\beta}$ is identically 0 on an neighborhood of K hence belongs also to D. Therefore

$$\varphi = \frac{1}{2}[2\ \Theta\circ\varphi + 2(\varphi - \Theta\circ\varphi)]$$

belongs to D.

The condition is certainly sufficient.

b) Let $\varphi \in C(X;E)$ be such that

$$\sup_{x \in K(D)} p(\varphi)^{\beta}(x) = r' < r.$$

Now we choose $r'' \in]r',r[$ and define $\psi : X \to E$ by

$$\psi(x) = \begin{cases} \varphi(x) & \text{if} \quad p(\varphi(x)) \leqslant r'' \\ \dfrac{r''}{p(\varphi(x))}\varphi(x) & \text{if} \quad p(\varphi(x)) \geqslant r''. \end{cases}$$

Of course ψ belongs to $C(X;E)$ and more precisely to $(r''/r)D$. As $(\varphi - \psi)^{\beta}$ is identically 0 on $K(D)$, we get $\varphi - \psi \in \varepsilon D$ for every $\varepsilon > 0$. For $\varepsilon > 0$ such that $r''/r + \varepsilon \leqslant 1$, we get then

$$\varphi = \psi + (\varphi - \psi) \in \frac{r''}{r}D + \varepsilon D \subset D.$$

Hence the conclusion.///

CRITERION II.1.5. _If_ D _is an absolutely convex and non void subset of_ $C(X;E)$, _then_ $x \in \beta X$ _belongs to_ $K(D)$ _if and only if, for every_

<u>neighborhood</u> V <u>of</u> x <u>in</u> βX, <u>there is</u> $\varphi \in C(X;E)\setminus D$ <u>such that</u> $\varphi^{\beta}(\beta X \setminus V) = \{0\}$.

<u>Proof</u>. The condition is necessary. If there is an open neighborhood V of x in βX such that φ belongs to D for every $\varphi \in C(X;E)$ verifying $\varphi^{\beta}(\beta X\setminus V) = \{0\}$, then $\beta X\setminus V$ is a hold of D and therefore x does not belong to K(D).

The condition is sufficient. If $x \in \beta X$ does not belong to K(D), there is a closed neighborhood V of x which is disjoint from K(D). Then $\beta X\setminus V$ is an open neighborhood of K(D) and every $\varphi \in C(X;E)$ such that $\varphi^{\beta}(\beta X\setminus V) = \{0\}$ belongs to D. Hence the conclusion.///

<u>II.2. LOCALIZATION OF THE SUPPORT</u> [20;5] & [28]

DEFINITIONS II.2.1.

a) Let us recall that a sequence $e_m \in E$ is <u>Mackey converging to</u> $e \in E$ if there is an absolutely convex and bounded subset B of E such that $e_m \to e$ in E_B. It is <u>fast converging to</u> $e \in E$ if there is an absolutely convex, bounded and completing subset (= a Banach disk) B of E such that $e_m \to e$ in E_B. Of course every fast converging sequence of E is Mackey converging in E to the same limit and every Mackey converging sequence of E is converging in E to the same limit.

In $C_{P}(X;E)$, one can also introduce the following stronger notion of convergence.

b) A sequence $\varphi_m \in C(X;E)$ is <u>superconverging to</u> $\varphi \in C(X;E)$ if there is an increasing sequence Ω_m of open subsets of βX which cover υX and are such that

$$(\varphi_m - \varphi)^{\beta}(\Omega_m) = \{0\}, \quad \forall\, m \in \mathbb{N}.$$

Of course every superconverging sequence of C(X;E) is converging in $C_{P}(X;E)$ and in $C_{Q,P}(X;E)$ to the same limit since, for every $B \in P$, $\bar{B}^{\upsilon X}$ is a compact subset of υX. It is possible to improve this information.

PROPOSITION II.2.2. <u>Every superconverging sequence of</u> $C_{P}(X;E)$ [resp. $C_{Q,P}(X;E)$] <u>is fast converging to the same limit</u>.

Proof. Let us prove the case $C_P(X;E)$; the other one is analogous.

Let φ_m be a sequence of $C_P(X;E)$ which is superconverging to 0. On one hand the sequence $m\varphi_m$ is also superconverging to 0 hence converges to 0. On the other hand, for every $\underline{c} \in l_1$, the series $\Sigma_{m=1}^{\infty} c_m m\varphi_m$ is certainly converging in $C_P(X;E)$. Therefore

$$K = \{ \sum_{m=1}^{\infty} c_m m\varphi_m : \sum_{m=1}^{\infty} |c_m| \leq 1\}$$

is an absolutely convex compact subset of $C_P(X;E)$ and it is trivial matter to check that the sequence φ_m converges to 0 in $C(X;E)_K$.///

LEMMA II.2.3. _If_ D _is an absolutely convex subset of_ C(X;E) _which absorbs every superconverging sequence and if_ Ω_m _is an increasing sequence of open subsets of_ βX _covering_ υX, _then there is_ $m \in \mathbb{N}$ _such that_ $K(D) \subset \overline{\Omega}_m^{\beta X}$.

Proof. If it is not the case, for every $m \in \mathbb{N}$, there is $\varphi_m \in C(X;E)\setminus D$ such that φ_m^{β} is identically 0 on a neighborhood of $\overline{\Omega}_m^{\beta X}$. Then the sequence $m\varphi_m$ is superconverging to 0 in C(X;E) and cannot be absorbed by D. Hence a contradiction.///

THEOREM II.2.4. _If_ D _is an absolutely convex subset of_ C(X;E) _which absorbs every superconverging sequence, then the support_ K(D) _of_ D _is contained in_ υX.

Proof. For every $x \in \beta X\setminus \upsilon X$, there is a bounded continuous function f on βX such that $f(y) > 0$ for every $y \in \upsilon X$ and $f(x) = 0$. Then the sets

$$\Omega_m = \{y \in \beta X : f(y) > \frac{1}{m}\}$$

constitute an increasing sequence of open subsets of βX covering υX. Hence the conclusion by the previous lemma since, for every $m \in \mathbb{N}$, we certainly habe $x \notin \overline{\Omega}_m^{\beta X}$.///

COROLLARY II.2.5. _If_ D _is an absolutely convex subset of_ $C_P(X;E)$ [resp. $C_{Q,P}(X;E)$] _which absorbs every absolutely convex and compact subset, then the support of_ D _is contained in_ υX.

Proof. It is a direct consequence of Proposition II.2.2.///

II.3. FINITE SUPPORT

THEOREM II.3.1. *If* D *is an absorbent absolutely convex subset of* C(X;E), *then the support* K(D) *of* D *is finite if and only if* D *absorbs every sequence* $\varphi_m \in C(X;E)$ *such that the sets*

$$\overline{\{x \in \beta X : \varphi_m^\beta(x) \neq 0\}}^{\beta X}$$

are mutually disjoint.

Proof. The condition is necessary. As K(D) is finite, there is at most a finite number of the sets

$$\overline{\{x \in \beta X : \varphi_m^\beta(x) \neq 0\}}^{\beta X}$$

which have a non void intersection with K(D). For the other values of m, we have

$$K(D) \subset \beta X \setminus \overline{\{x \in \beta X : \varphi_m^\beta(x) \neq 0\}}^{\beta X}$$

and therefore $\{x \in \beta X : \varphi_m^\beta(x) = 0\}$ is a neighborhood of K(D), which implies $\varphi_m \in D$. Hence the conclusion since D is absorbent.

The condition is sufficient. If it is not the case, K(D) contains a sequence x_m of distinct elements and, by the lemma II.11.6 of [20], we may suppose to have for each m a closed neighborhood V_m of x_m in βX, these neighborhoods being mutually disjoint. Then by the criterion II.1.5 there is a sequence $\varphi_m \in C(X;E) \setminus D$ such that

$$\varphi_m^\beta(\beta X \setminus V_m) = \{0\}.$$

Hence a contradiction since the sequence $m\varphi_m \in C(X;E)$ cannot be absorbed by D although the sets

$$\overline{\{x \in \beta X : (m\varphi_m)^\beta(x) \neq 0\}}^{\beta X}$$

are mutually disjoint.///

COROLLARY II.3.2. *If* D *is an absolutely convex subset of* $C_{Q,s}(X;E)$ *which absorbs the Mackey converging sequences, then* K(D) *is a finite subset of* υX.

Proof. By the theorem II.2.4 and the proposition II.2.2, we know already that $K(D)$ is a subset of υX.

Now let φ_m be a sequence of $C(X;E)$ such that the sets

$$\overline{\{x \in \beta X : \varphi_m^{\beta}(x) \neq 0\}}^{\beta X}$$

are mutually disjoint. Then the sequence $m\varphi_m$ is certainly bounded in $C_{Q,s}(X;E)$ and therefore the sequence φ_m is Mackey converging to 0 in that space hence absorbed by D. Hence the conclusion.///

II.4. BASIS OF AN ABSOLUTELY CONVEX SUBSET OF C(X;E)

DEFINITION II.4.1. The support $K(D)$ of an absolutely convex and non void subset D of $C(X;E)$ is a basis of D if every $\varphi \in C(X;E)$ such that $\varphi^{\beta}(K(D)) = \{0\}$ belongs to D.

Remark II.4.2. By the theorem II.1.4, we know already that the support of an absolutely convex and non void subset D of $C(X;E)$ is a basis of D if there are $p \in P$ and $r > 0$ such that

$$D \supset \{\varphi \in C(X;E) : p(\varphi(x)) \leqslant r, \forall x \in X\}.$$

THEOREM II.4.3. If D is an absolutely convex subset of $C_P(X;E)$ and if one of the following two conditions holds

a) D absorbs every Mackey converging sequence and E is metrizable,

b) D absorbs every fast converging sequence and E is a Fréchet space,

then $K(D)$ is a compact subset of υX and there are $p \in P$ and $r > 0$ such that

$$\{\varphi \in C(X;E) : p_{K(D)}(\varphi) \leqslant r\} \subset D;$$

in particular, $K(D)$ is a basis of D.

Proof. Let p_m be an increasing sequence of semi-norms on E such that $\{p_m : m \in \mathbb{N}\}$ is equivalent to P on E.

By the theorem II.2.4 and the proposition II.2.2, we know already that $K(D)$ is contained in υX.

To conclude by theorem II.1.4, it is then enough to prove that there are $m \in \mathbb{N}$ and $r > 0$ such that

$$\{\varphi \in C(X;E) : p_m(\varphi(x)) \leq r, \forall x \in X\} \subset D.$$

This we do by contradiction. Let us suppose the existence of a sequence $\varphi_m \in C(X;E) \setminus D$ such that

$$\sup_{x \in X} p_m(\varphi_m(x)) \leq m^{-3}.$$

a) Of course the sequence $m^2\varphi_m$ is bounded in $C_P(X;E)$ and the sequence $m\varphi_m$ is not absorbed by D. Hence a contradiction since the sequence $m\varphi_m$ converges to 0 in $C(X;E)_B$ where B is the absolutely convex hull of the sequence $m^2\varphi_m$, an absolutely convex and bounded subset of $C_P(X;E)$ indeed.

b) Of course the sequence $m^2\varphi_m$ converges to 0 in $C_P(X;E)$. Moreover for every $\underline{c} \in l_1$, the series $\Sigma_{m=1}^{\infty} c_m m^2 \varphi_m$ converges in $C_P(X;E)$: infact it even converges uniformly on X since for $k \leq r \leq s$ we have

$$\sup_{x \in X} p_k\left(\sum_{m=r}^{s} c_m m^2 \varphi_m\right) \leq \sum_{m=r}^{s} |c_m|.$$

Therefore

$$K = \left\{\sum_{m=1}^{\infty} c_m m^2 \varphi_m : \sum_{m=1}^{\infty} |c_m| \leq 1\right\}$$

is an absolutely convex and compact subset of $C_P(X;E)$. Hence a contradiction since the sequence $m\varphi_m$ is not absorbed by D but converges to 0 in $C(X;E)_K$.///

An analogous proof leads directly to the following result.

THEOREM II.4.3'. *If* D *is an absolutely convex subset of* $C_{Q,P}(X;E)$ *and if one of the following two conditions holds*

a) D *absorbs every Mackey converging sequence and* (E,Q) *is metrizable,*

b) D *absorbs every fast converging sequence and* (E,Q) *is a Fréchet space,*

then K(D) *is a compact subset of* υX *and there are* $q \in Q$ *and* $r > 0$

<u>such that</u>

$$\{\varphi \in C(X;E) : \sup_{x\in X} q(\varphi(x)) \leq r\} \subset D.///$$

The particular case of the theorem II.4.3 can be improved in different ways.

PROPOSITION II.4.4. <u>If</u> E <u>is metrizable and if the absolutely convex subset</u> D <u>of</u> $C_P(X;E)$ <u>absorbs the fast converging sequences, then</u> $K(D)$ <u>is a compact subset of</u> υX <u>and a basis of</u> D.

<u>Proof</u>. Let p_m be an increasing sequence of semi-norms on E such that $\{p_m : m \in \mathbb{N}\}$ is equivalent to P on E and let φ be an element of $C(X;E)$ such that $\varphi^{\beta}(K(D)) = \{0\}$.

By the theorem II.2.4 and the proposition II.2.2, we know already that $K(D)$ is contained in υX. The sets

$$\Omega_m = \{x \in \upsilon X : p_m(\varphi(x)) < m^{-3}\} \quad (m \in \mathbb{N})$$

are decreasing open subsets of υX which intersection contains $K(D)$. For every m, let us fix $f_m \in C(X)$ with values in $[0,1]$, identically 1 on a neighorhood of $K(D)$ and 0 on $\upsilon X \setminus \Omega_m$.

On one hand the sequence $m^2 f_m \varphi$ converges of course to 0 in $C_P(X;E)$. On the other hand for every $\underline{c} \in l_1$, the series $\Sigma_{m=1}^{\infty} c_m m^2 f_m \varphi$ is uniformly Cauchy on X and converges in E at every $x \in X$ since at every $x \in \cap_{m=1}^{\infty} \Omega_m$ all its terms are 0 and at every $x \in X \setminus \cap_{m=1}^{\infty} \Omega_m$ it reduces to a finite sum; hence the series $\Sigma_{m=1}^{\infty} c_m m^2 f_m \varphi$ converges in $C_P(X;E)$.

Therefore

$$K = \{ \sum_{m=1}^{\infty} c_m m^2 f_m \varphi : \sum_{m=1}^{\infty} |c_m| \leq 1\}$$

is an absolutely convex compact subset of $C_P(X;E)$. As the sequence $m f_m \varphi$ converges to 0 in $C(X;E)_K$, it is absorbed by D : there is $C > 0$ such that $f_m \varphi \in (C/m)D$ for every m.

For every m, $[2(1-f_m)\varphi]^{\beta}$ is identically 0 on a neighborhood of $K(D)$ hence $(1-f_m)\varphi$ belongs to $(1/2)D$.

Therefore for m such that $C/m \leqslant 1/2$ we get

$$\varphi = f_m \varphi + (1-f_m)\varphi \in D$$

hence the conclusion.///

PROPOSITION II.4.5 [6] If P is contained in $K(X)$ (which is always the case if X is realcompact) [resp. If E is realcompact] and if there is an increasing sequence $(q_m)_{m \in \mathbb{N}}$ of continuous semi-norms on E such that, for every $\varphi \in C(X;E)$ and $K \in P$, $\{q_m : m \in \mathbb{N}\}$ is equivalent to P on the linear hull $>\varphi(K)<$ of $\varphi(K)$, then the support of every absolutely convex and bornivorous subset D of $C_P(X;E)$ is contained in υX and is a basis of D.

Proof. By the theorem II.2.4 and the proposition II.2.2, we know already that $K(D)$ is contained in υX.

If E is realcompact, the spaces $C_P(X;E)$ and $C_P(\upsilon X;E)$ are canonically isomorphic. Therefore without loss of generality we may suppose that P is contained in $K(X)$.

Now let us consider an element φ of $C(X;E)$ such that $\varphi^\beta(K(D)) = \{0\}$.

For every $m \in \mathbb{N}$, let us define $\Theta_m : E \to E$ by

$$\Theta_m(e) = \begin{cases} e & \text{if} \quad q_m(e) \leqslant 1/(2m^2) \\ (2-2m^2 q_m(e))e & \text{if} \quad 1/(2m^2) \leqslant q_m(e) \leqslant 1/m^2 \\ 0 & \text{if} \quad 1/m^2 \leqslant q_m(e) \end{cases}$$

It is direct matter to check that Θ_m is continuous and such that

$$q_m(\Theta_m(e)) \leqslant \frac{1}{2m^2}, \quad \forall\, e \in E.$$

On one hand, for every $m \in \mathbb{N}$,

$$U_m = \{e \in \beta E : (q_m)^\beta(e) < \frac{1}{2m^2}\}$$

is an open subset of βE containing $\varphi^\beta(K(D))$. As Θ_m reduces to the identity on $U_m \cap E$, we get that $(\Theta_m)^\beta$ is the identity map on U_m. Therefore $(\varphi^\beta)^{-1}(U_m)$ is an open neighborhood of $K(D)$ in βX where φ^β and

$(\Theta_m o\varphi)^\beta$ coincide. Hence we get

$$\Theta_m o\varphi - \varphi \in rD, \quad \forall\, r > 0.$$

On the other hand, for every $K \in P$ and $m \in \mathbb{N}$, we have

$$m\, \Theta_m o\varphi(K) \subset \{e \in {>}\varphi(K){<} : q_m(e) \leq \frac{1}{2m}\}.$$

As $m\, \Theta_m o\varphi(K)$ is a compact subset of E for every $m \in \mathbb{N}$, this implies that

$$\bigcup_{m=1}^{\infty} m\, \Theta_m o\varphi(K)$$

is a bounded subset of $>\varphi(K)<$ hence of E for every $K \in P$. Therefore the sequence $m\, \Theta_m o\varphi$ is bounded in $C_P(X;E)$ hence is absorbed by D : there is $C > 0$ such that $m\, \Theta_m o\varphi \in CD$ for every $m \in \mathbb{N}$.

For $m \in \mathbb{N}$ large enough so that $C/m \leq 1/2$, we get

$$\varphi = \Theta_m o\varphi - (\Theta_m o\varphi - \varphi) \in \tfrac{1}{2}D + \tfrac{1}{2}D = D.$$

Hence the conclusion.///

PROPOSITION II.4.6. <u>If</u> D <u>is an absolutely convex subset of</u> $C_{Q,s}(X;E)$ <u>which absorbs the Mackey converging sequences and if</u> X <u>is realcompact and verifies the first axiom of countability, then</u> K(D) <u>is a basis of</u> D.

<u>Proof</u>. By corollary II.3.2, we know that K(D) is a finite subset $\{x_1, \ldots, x_J\}$ of υX. By hypothesis each one of the points x_j $(j \leq J)$ has a countable basis of neighborhoods, let us say $\{\Omega_{j,m} : m \in \mathbb{N}\}$. Therefore $\Omega_m = \cup_{j=1}^{J} \Omega_{j,m}$ is a fundamental sequence of neighborhoods of K(D). Now let us choose $f_1 \in C(X)$ identically 0 on $X \setminus \Omega_1$ and 1 on a neighborhood V_1 of K(D). By recursion we can then choose functions $f_m \in C(X)$ $(m \geq 2)$ such that each f_m is identically 0 on $X \setminus (\Omega_m \cap V_{m-1})$ and 1 on a neighborhood V_m of K(D).

Let $\varphi \in C(X;E)$ be such that $\varphi^\beta(K(D)) = \{0\}$. Of course the sequence $m^2 f_m \varphi$ converges to 0 in $C_{Q,s}(X;E)$. Therefore the sequence $m f_m \varphi$ converges to 0 in $C(X;E)_B$ where B is the absolutely convex hull of the sequence $m^2 f_m \varphi$, an absolutely convex bounded subset of $C_{Q,s}(X;E)$ indeed. Hence the sequence $mf_m\varphi$ is absorbed by D.

Moreover for every $m \in \mathbb{N}$, $[(1-f_m)\varphi]^\beta$ is identically 0 on a neighborhood of K(D) therefore $2(1-f_m)\varphi$ belongs to D.

Hence for $m \in \mathbb{N}$ large enough, $\varphi = f_m\varphi + (1-f_m)\varphi$ belongs to D. Hence the conclusion.///

II.5. SUPPORT OF AN ABSOLUTELY CONVEX BORNIVOROUS SUBSET

LEMMA II.5.1. Given a subset A of βX, the following conditions are equivalent

a) if the sequence φ_m converges to 0 in $C_P(X;E)$, then one has

$$\sup_{x\in A} p(\varphi_m)^\beta(x) \to 0, \quad \forall\, p \in P,$$

b) for every bounded subset B of $C_P(X;E)$, one has

$$\sup_{x\in A}\sup_{\varphi\in B} p(\varphi)^\beta(x) < +\infty, \quad \forall\, p \in P,$$

c) every P-finite sequence Ω_m of open subsets of βX is A-finite (i.e. if $\{m : \Omega_m \cap B \neq \emptyset\}$ is finite for every $B \in P$, then $\{m : \Omega_m \cap A \neq \emptyset\}$ is finite),

and they imply that A is contained in υX.

Proof. (a) ⇒ (b) is standard.

(b) ⇒ (c) The proof is analogous to the one of lemma II.11.1 of [20] and is omitted.

(c) ⇒ (a) If this is not the case, there is a sequence φ_m converging to 0 in $C_P(X;E)$, $p \in P$ and $r > 0$ such that

$$\sup_{x\in A} p(\varphi_m)^\beta(x) \geq r, \quad \forall\, m.$$

Then

$$\{x \in \beta X : p(\varphi_m)^\beta(x) > \tfrac{r}{2}\} \quad (m \in \mathbb{N})$$

is a P-finite sequence of open subsets of βX which is not A-finite hence a contradiction.

The conclusion is then a direct consequence of the fact that for every $x_o \in \beta X \setminus \upsilon X$, there is $f \in C(X)$ with values in $[0,1]$ such that $f(y) > 0$ for every $y \in \upsilon X$ and $f^\beta(x_o) = 0$. In fact

$$V_m = \{x \in \beta X : f^\beta(x) > \frac{1}{m}\} \quad (m \in \mathbb{N})$$

is then an increasing sequence of open subsets of βX which cover υX. As every $B \in P$ is relatively compact in υX, this implies $B \subset V_m$ for m large enough. Therefore the sequence

$$U_m = \{x \in \beta X : f^\beta(x) < \frac{1}{m}\} \quad (m \in \mathbb{N})$$

is P-finite and decreasing hence A-finite, which implies $x_o \notin A$.///

In the same manner one gets the following result.

LEMMA II.5.1'. Given a subset A of X, the following conditions are equivalent

a) if the sequence φ_m converges to 0 in $C_{Q,P}(X;E)$, then one has

$$\sup_{m \in \mathbb{N}} \sup_{x \in A} q(\varphi_m(x)) < +\infty, \quad \forall\, q \in Q,$$

b) for every bounded subset B of $C_{Q,P}(X;E)$, one has

$$\sup_{x \in A} \sup_{\varphi \in B} q(\varphi(x)) < +\infty, \quad \forall\, q \in Q,$$

and they imply that every P-finite sequence of open subsets of X is A-finite.

Conversely if every P-finite sequence of subsets of X of the type $\{x \in X : q(\varphi_m(x)) > r_m\}$ with $q \in Q$, $\varphi_m \in C(X;E)$ and $r_m > 0$ is A-finite, then one has

$$\sup_{x \in A} q(\psi_m(x)) \to 0$$

for every $q \in Q$ and every sequence ψ_m converging to 0 in $C_{Q,P}(X;E)$.///

PROPOSITION II.5.2. If D is an absolutely convex and bornivorous subset of $C_P(X;E)$, then K(D) is contained in υX and is such that

$$\sup_{\varphi \in B} p_{K(D)}(\varphi) < +\infty \qquad (\star)$$

<u>for every</u> $p \in P$ <u>and every bounded subset</u> B <u>of</u> $C_P(X;E)$. <u>In particular</u>, $K(D)$ <u>belongs</u> $\tilde{P}^{\upsilon}$.

<u>Proof</u>. By corollary II.2.5, we know already that $K(D)$ is contained in υX.

Let us establish the inequality $(\star)$ by contradiction. If it is not the case, by lemma II.5.1, there are a P-finite sequence Ω_m of open subsets of υX and a sequence $x_m \in \Omega_m \cap K(D)$. By the criterion II.1.5, there is then a sequence $\varphi_m \in C(X;E) \setminus D$ such that

$$(\varphi_m)^{\beta}(\upsilon X \setminus \Omega_m) = \{0\}$$

for every $m \in \mathbb{N}$. Then the sequence $m\varphi_m$ cannot be absorbed by D although it is bounded in $C_P(X;E)$. Hence a contradiction.

The particular case is then immediate since, for every bounded subset B of $C_P(X)$ and every $e \in E$, $\{fe : f \in B\}$ is a bounded subset of $C_P(X;E)$.///

II.6. THE SUPPORT OF A LINEAR FUNCTIONAL ON C(X;E)

DEFINITION II.6.1. For every linear functional φ' on $C(X;E)$,

$$D_{\varphi'} = \{\varphi \in C(X;E) : |\langle\varphi,\varphi'\rangle| \leq 1\}$$

is of course an absolutely convex and non void subset of $C(X;E)$. Therefore one can consider the support $K(D_{\varphi'})$ of $D_{\varphi'}$ that we shall rather call the <u>support of</u> φ' and designate by the notation

$$\operatorname{supp} \varphi'.$$

The foregoing results lead immediately to the following statements.

<u>Remarks</u> II.6.2. Let φ' be a linear functional on $C(X;E)$.

a) If there are $p \in P$ and $r > 0$ such that

$$\varphi \in C(X;E) \ \& \ \sup_{x\in X} p(\varphi(x)) \leq r \quad \Rightarrow \quad |\langle\varphi,\varphi'\rangle| \leq 1,$$

then supp φ' is the smallest compact subset K of βX such that $\langle\varphi,\varphi'\rangle = 0$ for every $\varphi \in C(X;E)$ such that φ^{β} is identically 0 on K. Moreover we have then

$$\varphi \in C(X;E) \ \& \ \sup_{x\in \mathrm{supp}\,\varphi'} p(\varphi)^{\beta}(x) \leq r \quad \Rightarrow \quad |\langle\varphi,\varphi'\rangle| \leq 1.$$

(Cf. Theorem II.1.4)

b) If φ' is bounded on every superconverging sequence [in particular, if φ' is bounded on every absolutely convex and compact subset or on every bounded subset of $C_{P}(X;E)$ or of $C_{Q,P}(X;E)$], then supp φ' is a compact subset of υX. (Cf. Theorem II.2.4)

c) If φ' is bounded on every Mackey converging sequence of $C_{Q,s}(X;E)$, then supp φ' is a finite subset of υX. (Cf. corollary II.3.2) If moreover X is realcompact and verifies the first axiom of countability, then we have $\langle\varphi,\varphi'\rangle = 0$ for every $\varphi \in C(X;E)$ such that $\varphi^{\upsilon}(\mathrm{supp}\,\varphi') = \{0\}$. (Cf. proposition II.4.6)

d) If φ' is bounded on every Mackey converging sequence [resp. on every fast converging sequence] of $C_{P}(X;E)$ and if E is metrizable [resp. is a Fréchet space], then supp φ' is a compact subset of υX and there are $p \in P$ and $C > 0$ such that

$$|\langle\varphi,\varphi'\rangle| \leq C\, p_{\mathrm{supp}\,\varphi'}(\varphi), \quad \forall\, \varphi \in C_{P}(X;E).$$

(Cf. Theorem II.4.3)

e) If φ' is bounded on every fast converging sequence of $C_{P}(X;E)$ and if E is metrizable, then supp φ' is a compact subset of υX and we have $\langle\varphi,\varphi'\rangle = 0$ for every $\varphi \in C(X;E)$ such that $\varphi^{\upsilon}(\mathrm{supp}\,\varphi') = \{0\}$. (Cf. proposition II.4.4)

For the bounded linear functionals, we already have the following property.

PROPOSITION II.6.3. <u>If</u> φ' <u>is a bounded linear functional on</u> $C_{P}(X;E)$, <u>then</u> supp φ' <u>belongs to</u> $\tilde{P}^{\upsilon}$. <u>If moreover</u> E <u>is metrizable, the-</u>

re are $p \in P$ and $C > 0$ such that

$$|\langle .,\varphi' \rangle| \leq C\, p_{\mathrm{supp}\,\varphi'}(.) \text{ on } C_{\mathcal{P}}(X;E).$$

Proof. This is a direct consequence of the proposition II.5.2 and of the theorem II.4.3.///

Let us note also that the propositionsII.4.5 and II.5.2 lead to the following result : if $\mathcal{P}$ is contained in $\mathcal{K}(X)$ [resp. If E is real-compact] and if there is an increasing sequence $(q_m)_{m\in\mathbb{N}}$ of continuous semi-norms on E such that, for every $\varphi \in C(X;E)$ and $K \in \mathcal{P}$, $\{q_m : m\in\mathbb{N}\}$ is equivalent to P on the linear hull $>\varphi(K)<$ of $\varphi(K)$, then, for every bounded linear functional φ' on $C_{\mathcal{P}}(X;E)$, supp φ' belongs to $\tilde{\mathcal{P}}^{\upsilon}$ and we have $\langle \varphi,\varphi' \rangle = 0$ for every $\varphi \in C(X;E)$ such that $\varphi^{\upsilon}(\mathrm{supp}\,\varphi') = \{0\}$.

Now let us come to the case of the continuous linear functionals.

PROPOSITION II.6.4. For every continuous linear functional φ' on $C_{\mathcal{P}}(X;E)$, there are $p \in P$, $B \in \mathcal{P}$ and $C > 0$ such that supp $\varphi' \subset \bar{B}^{\upsilon X}$ and

$$|\langle .,\varphi' \rangle| \leq C\, p_{\mathrm{supp}\,\varphi'}(.). \qquad (\star)$$

Moreover supp φ' is the smallest compact subset of υX for which there are $p \in \mathcal{P}$ and $C > 0$ such that the inequality $(\star)$ holds.

Proof. By hypothesis there are $p \in P$, $B \in \mathcal{P}$ and $C > 0$ such that $|\langle .,\varphi' \rangle| \leq C\, p_B(.)$ hence such that

$$\varphi \in C(X;E) \;\&\; \sup_{x\in X} p(\varphi(x)) \leq \frac{1}{C} \quad \Rightarrow \quad |\langle \varphi,\varphi' \rangle| \leq 1.$$

From this, we get easily the inclusions supp $\varphi' \subset \bar{B}^{\upsilon X}$ and

$$\{\varphi \in C(X;E) : p_{\mathrm{supp}\,\varphi'}(\varphi) < \tfrac{1}{C}\} \subset D_{\varphi'}.$$

The conclusion then follows at once.///

The same method leads to the following result.

PROPOSITION II.6.4'. For every continuous linear functional φ' on $C_{Q,\mathcal{P}}(X;E)$, there is $B \in \mathcal{P}$ such that supp $\varphi' \subset \bar{B}^{\upsilon X}$.///

II.7. STRUCTURE OF THE DUAL OF $C_{Q,s}(X;E)$

In this paragraph we represent the continuous linear functionals on $C_{Q,s}(X;E)$ and establish a few properties of the support of certain classes of subsets of $C_{Q,s}(X;E)'$ which will serve us as a guide in chapter III.

THEOREM II.7.1. *If* φ' *is a linear functional on* $C(X;E)$ *for which there are* $q \in Q$, $A \in \mathcal{A}(X)$ *and* $C > 0$ *such that*

$$|\langle\varphi,\varphi'\rangle| \leqslant C\, q_A(\varphi), \quad \forall\, \varphi \in C(X;E),$$

then there are $e'_x \in (E,Q)'$ $(x \in \mathrm{supp}\,\varphi')$ *such that*

$$|\langle e,e'_x\rangle| \leqslant C\, q(e), \quad \forall\, e \in E,$$

for which we have

$$\langle\varphi,\varphi'\rangle = \sum_{x\in \mathrm{supp}\,\varphi'} \langle\varphi(x),e'_x\rangle, \quad \forall\, \varphi \in C(X;E),$$

this representation being unique.

Conversely for every $A \in \mathcal{A}(X)$ *and* $e'_x \in (E,Q)'$ $(x \in A)$,

$$\sum_{x\in A} \langle .(x),e'_x\rangle$$

is a continuous linear functional on $C_{Q,s}(X;E)$.

Proof. Let φ' be a linear functional on $C(X;E)$ such that

$$|\langle\varphi,\varphi'\rangle| \leqslant C\, q_A(\varphi), \quad \forall\, \varphi \in C(X;E), \qquad (\star)$$

with $A \in \mathcal{A}(X)$, $C > 0$ and $q \in Q$. Of course we have $\mathrm{supp}\,\varphi' \subset A$. As A is finite, there are functions $f_x \in C(X)$ $(x \in A)$ such that $f_x(x) = 1$ and which supports are mutually disjoint. Then for every $\varphi \in C(X;E)$, each of the $f_x\varphi$ belongs to $C(X;E)$ and therefore we have

$$\langle\varphi,\varphi'\rangle = \langle \sum_{x\in A} f_x\varphi,\varphi'\rangle + \langle\varphi - \sum_{x\in A} f_x\varphi,\varphi'\rangle$$

$$= \sum_{x\in A} \langle f_x\varphi,\varphi'\rangle$$

by use of the relation (*). Now let us remark that for every $\varphi \in C(X;E)$, $<f_x\varphi,\varphi'>$ depends only on $\varphi(x)$: if $\varphi,\psi \in C(X;E)$ are such that $\varphi(x) = \psi(x)$, we have $q_A(f_x\varphi - f_x\psi) = 0$ hence

$$<f_x\varphi,\varphi'> = <f_x\psi,\varphi'>.$$

Therefore for every $x \in A$, we can definie a functional e'_x on E by

$$<e,e'_x> = <f_x e,\varphi'>, \quad \forall\, e \in E;$$

it is of course a linear functional on E and it is continuous on (E,Q) since we have

$$|<e,e'_x>| = |<f_x e,\varphi'>| \leq C\, q_A(f_x e) = C\, q(e), \quad \forall\, e \in E.$$

Moreover let us remark that for $x \in A \setminus \text{supp } \varphi'$, $f_x e$ is identically 0 on a neighborhood of supp φ' for every $e \in E$ which implies $e'_x = 0$.

The uniqueness of the representation and the converse property are immediate.///

DEFINITION II.7.2. By use of the uniqueness of this representation of the continuous linear functionals φ' on $C_{Q,s}(X;E)$, we may adopt the precise notation

$$e'_{\varphi',x} \quad (x \in \text{supp } \varphi')$$

for the unique functionals e'_x ($x \in \text{supp } \varphi'$). In this way, the unique representation of every $\varphi' \in C_{Q,s}(X;E)'$ is given by

$$<.,\varphi'> = \sum_{x \in \text{supp } \varphi'} <.(x),e'_{\varphi',x}>.$$

DEFINITION II.7.3. The <u>support</u> of a subset B of $C_{Q,s}(X;E)'$ is denoted by

$$\text{supp } B$$

and designates the set

$$\text{supp } B = \cup \{\text{supp } \varphi' : \varphi' \in B\}.$$

PROPOSITION II.7.4. For every bounded subset B of $C_{Q,s}(X;E)'_s$, supp B is a bounding subset of X.

Proof. Let B be a subset of $C_{Q,s}(X;E)'_s$ such that supp B is not a bounding subset of X. There is then $f \in C(X)$ which is unbounded on supp B and by use of proposition II.11.9 of [20], a sequence $x_m \in$ supp B and a sequence $f_m \in C(X)$ such that $f_m(x_m) = 1$ for every $m \in \mathbb{N}$ and that the supports supp f_m are mutually disjoint and locally finite. There is then a sequence $\varphi'_m \in B$ such that $x_m \in$ supp φ'_m for every m and since every one of the sets supp φ'_m is finite, up to going to a subsequence, we may suppose that we have

$$\text{supp } f_{m+1} \cap \left(\bigcup_{k=1}^{m} \text{supp } \varphi'_k\right) = \emptyset, \quad \forall\, m \in \mathbb{N}.$$

One can then easily get sequences $g_m \in C(X)$, $e_m \in E$ and $c_m \in \mathbb{C}$ such that

$$\text{supp } g_m \subset \text{supp } f_m,$$

$$g_m(x) = \delta_{x,x_m}, \quad \forall\, x \in \text{supp } \varphi'_m,$$

and

$$<\sum_{k=1}^{m} c_k g_k e_k, \varphi'_m> = m$$

for every $m \in \mathbb{N}$. Now the series $\Sigma_{k=1}^{\infty} c_k g_k e_k$ converges in $C_{Q,s}(X;E)$ [the functions g_m have supports which are mutually disjoint and locally finite] and is certainly such that

$$<\sum_{k=1}^{\infty} c_k g_k e_k, \varphi'_m> = <\sum_{k=1}^{m} c_k g_k e_k, \varphi'_m> = m, \quad \forall\, m \in \mathbb{N},$$

which implies that B is not bounded in $C_{Q,s}(X;E)'_s$. Hence the conclusion.///

PROPOSITION II.7.5. A subset B of $C_{Q,s}(X;E)'_b$ is bounded if and only if supp B is finite and such that, for every $x \in B$, $\{e'_{\varphi',x} : \varphi' \in B\}$ is bounded in $(E,Q)'_b$.

Proof. The condition is necessary.

On one hand we establish that if $B \subset C_{Q,s}(X;E)'_b$ has not a finite support, then B is not bounded. In fact by lemma II.11.6 of [20], there is then a sequence $x_m \in \text{supp } B$ and a sequence V_m of closed and mutually disjoint neighborhoods of the x_m. For every $m \in \mathbb{N}$, there is then $\varphi'_m \in B$ such that $e'_{\varphi'_m, x_m}$ differs from 0 and, up to going to a subsequence, we may ask that, for every $m \in \mathbb{N}$, x_{m+1} does not belong to $\cup_{k=1}^m \text{supp } \varphi'_k$. By recursion, one may then get sequences $f_m \in C(X)$ and $e_m \in E$ such that

$$\text{supp } f_m \subset V_m, \; f_m(x_m) = 1 \quad \text{and} \quad \langle f_m e_m, \varphi'_m \rangle = m$$

for every $m \in \mathbb{N}$. As the f_m have mutually disjoint supports, the sequence $f_m e_m$ is bounded in $C_{Q,s}(X;E)$ and therefore B is not bounded in $C_{Q,s}(X;E)'_b$.

On the other hand let us establish that if B is a bounded subset of $C_{Q,s}(X;E)'_b$ then, for every $x \in \text{supp } B$, $\{e'_{\varphi',x} : \varphi' \in B\}$ is a bounded subset of $(E,Q)'_b$. As supp B is finite, there are functions $f_x \in C(X)$ $(x \in \text{supp } B)$ such that $f_x(y) = \delta_{x,y}$ for every $x,y \in \text{supp } B$. To conclude, it is enough then to notice that for every bounded subset B of (E,Q) and every $x \in \text{supp } B$, the set

$$B_x = \{f_x e : e \in B\}$$

is bounded in $C_{Q,s}(X;E)$ hence such that

$$\sup_{e \in B} \sup_{\varphi' \in B} |\langle e, e'_{\varphi',x} \rangle| = \sup_{\varphi' \in B} \sup_{\varphi \in B_x} |\langle \varphi, \varphi' \rangle| < +\infty.$$

The sufficiency of the condition is immediate.///

PROPOSITION II.7.6. <u>A subset</u> B <u>of</u> $C_{Q,s}(X;E)'$ <u>is equicontinuous if and only if</u> supp B <u>is finite and such that, for every</u> $x \in \text{supp } B$, $\{e'_{\varphi',x} : \varphi' \in B\}$ <u>is an equicontinuous subset of</u> $(E,Q)'$.

<u>Proof</u>. The necessity of the condition is a direct consequence of theorem II.7.1; its sufficiency is immediate.///

CHAPTER III

STRUCTURE OF THE DUAL OF $C_p(X;E)$

III.1. SCALAR BOREL MEASURES

Our aim in this paragraph is not to develop a theory of the scalar Borel measures on X but rather to set up the vocabulary and get the results that we use later on.

DEFINITION III.1.1. We denote by

$$\underline{B}(X)$$

the smallest subset of $p(X)$ containing all the open subsets of X, the complement in X of its elements and the countable unions of its elements. Therefrom we get immediately that $\underline{B}(X)$ contains all the countable intersections of its elements.

A Borel subset of X is any element of $\underline{B}(X)$. A Borel partition of $b \in \underline{B}(X)$ is a partition $\{b_\alpha : \alpha \in A\}$ of b such that $b_\alpha \in \underline{B}(X)$ for every $\alpha \in A$.

DEFINITIONS III.1.2. A scalar Borel measure on X is a function

$$\mu : \underline{B}(X) \to \mathbb{C}.$$

The scalar Borel measure μ on X

a) is finitely additive if we have $\mu(\emptyset) = 0$ and if, for every $b \in \underline{B}(X)$ and every finite Borel partition $\{b_1, \ldots, b_J\}$ of b, we have

$$\mu(b) = \sum_{j=1}^{J} \mu(b_j).$$

b) is <u>countably additive</u> if it is finitely additive and if, for every $b \in \underline{B}(X)$ and every infinite countable Borel partition $\{b_j : j \in \mathbb{N}\}$ of b, we have

$$\mu(b) = \sum_{j=1}^{\infty} \mu(b_j),$$

the series converging in $\mathbb{C}$. This implies of course that the series $\Sigma_{j=1}^{\infty} \mu(b_j)$ is absolutely converging in $\mathbb{C}$.

c) has a <u>finite variation on</u> $b \in \underline{B}(X)$ if we have

$$\sup_{P \in P(b)} \sum_{b' \in P} |\mu(b')| < \infty$$

where $P(b)$ is the set of all the finite Borel partitions of b.

d) has a <u>finite variation</u> if it has a finite variation on every $b \in \underline{B}(X)$. This is certainly the case if it has a finite variation on X.

e) is <u>regular</u> if, for every $b \in \underline{B}(X)$ and every $\varepsilon > 0$, there are a closed subset F and an open subset Ω of X such that $F \subset b \subset \Omega$ and that for every $b' \in \underline{B}(X)$ contained in $\Omega \setminus F$, we have $|\mu(b')| \leqslant \varepsilon$.

NOTATION III.1.3. If the scalar Borel measure μ on X has a finite variation on $b \in \underline{B}(X)$, we set

$$V\mu(b) = \sup_{P \in P(b)} \sum_{b' \in P} |\mu(b')|$$

where $P(b)$ is the set of all the finite Borel partitions of b.

In this way if the scalar Borel measure μ on X has a finite variation, $V\mu$ appears as a scalar Borel measure on X with values in $[0,+\infty[$. This can be improved a lot.

PROPOSITION III.1.4. <u>If the scalar Borel measure</u> μ <u>on</u> X <u>is finitely additive and has a finite variation, then the scalar Borel measure</u> $V\mu$ <u>on</u> X <u>is finitely additive and has a finite variation</u>.

a) <u>If moreover</u> μ <u>is countably additive, then</u> $V\mu$ <u>is countably additive</u>.

b) <u>If moreover</u> μ <u>is regular, then</u> $V\mu$ <u>is regular</u>.

<u>Proof</u>. Let us prove first that $V\mu$ is finitely additive. Let b be

a Borel subset of X and let P' be a finite Borel partition of b. The inequality

$$\sum_{b' \in P'} V\mu(b') \leq V\mu(b)$$

is a direct consequence of the fact that for finite Borel partitions $P(b')$ of each of the $b' \in P'$,

$$\cup \{P(b') : b' \in P'\}$$

is a finite Borel partition of b. Conversely for every finite Borel partition P'' of b and every $b' \in P'$,

$$\{b' \cap b'' \neq \emptyset : b'' \in P''\}$$

is a finite Borel partition of b'. Therefore we get

$$\sum_{b'' \in P''} |\mu(b'')| \leq \sum_{b' \in P'} \sum_{b'' \in P''} |\mu(b' \cap b'')| \leq \sum_{b' \in P'} V\mu(b')$$

hence the other inequality.

Therefore $V\mu$ has a finite variation since every finitely additive scalar Borel measure μ with values in $[0,+\infty[$ has of course a finite variation.

a) Let b be a Borel subset of X and let $\{b_m : m \in \mathbb{N}\}$ be an infinite countable Borel partition of b. On one hand, for every $M \in \mathbb{N}$, we have

$$\sum_{m=1}^{M} V\mu(b_m) + V\mu(b \setminus \bigcup_{m=1}^{M} b_m) = V\mu(b)$$

hence

$$\sum_{m=1}^{\infty} V\mu(b_m) \leq V\mu(b).$$

On the other hand for every $\varepsilon > 0$, there is a finite Borel partition $\{b'_1, \ldots, b'_J\}$ of b such that

$$V\mu(b) \leq \sum_{j=1}^{J} |\mu(b'_j)| + \varepsilon$$

hence such that

$$V\mu(b) \underset{(\star)}{\leq} \sum_{j=1}^{J} \left| \sum_{m=1}^{\infty} \mu(b_m \cap b'_j) \right| + \varepsilon$$

$$\leq \sum_{m=1}^{\infty} \sum_{j=1}^{J} |\mu(b_m \cap b'_j)| + \varepsilon$$

$$\underset{(\star\star)}{\leq} \sum_{m=1}^{\infty} V\mu(b_m) + \varepsilon$$

[in $(\star)$, we use the fact that μ is countably additive; in $(\star\star)$, we use the definition of $V\mu$]; hence the other inequality

$$V\mu(b) \leq \sum_{m=1}^{\infty} V\mu(b_m).$$

b) To establish this, it is of course enough to prove that if $b \in \underline{B}(X)$, the open subset Ω of X and the closed subset F of X are such that $F \subset b \subset \Omega$ and that we have $|\mu(b')| \leq \varepsilon$ for every Borel subset b' of X contained in $\Omega \setminus F$, then we also have $V\mu(b') \leq 4\varepsilon$ for every $b' \in \underline{B}(X)$ contained in $\Omega \setminus F$. To get this, it is enough to note that for every finite Borel partition P of such a set b', we successively have [for $c \in \mathbb{C}$, Rc is the real part of c and Ic the imaginary part of c; for $r \in \mathbb{R}$, r_+ is the positive part of r and r_- the negative part of r]

$$\sum_{b''\in P} |\mu(b'')| \leq \sum_{\substack{b''\in P \\ R\mu(b'')\geq 0}} [R\mu(b'')]_+ + \sum_{\substack{b''\in P \\ R\mu(b'')<0}} [R\mu(b'')]_-$$

$$+ \sum_{\substack{b''\in P \\ I\mu(b'')\geq 0}} [I\mu(b'')]_+ + \sum_{\substack{b''\in P \\ I\mu(b'')<0}} [I\mu(b'')]_-$$

$$\leq \Big|\mu\Big(\bigcup_{\substack{b''\in P \\ R\mu(b'')\geq 0}} b''\Big)\Big| + \Big|\mu\Big(\bigcup_{\substack{b''\in P \\ R\mu(b'')<0}} b''\Big)\Big|$$

$$+ \Big|\mu\Big(\bigcup_{\substack{b''\in P \\ I\mu(b'')\geq 0}} b''\Big)\Big| + \Big|\mu\Big(\bigcup_{\substack{b''\in P \\ I\mu(b'')<0}} b''\Big)\Big| \leq 4\varepsilon.$$

Hence the conclusion.///

III.2. VECTOR-VALUED BOREL MEASURES

Our aim in this paragraph is not to develop a theory of the vector-valued Borel measures on X but rather to set up the vocabulary and get the results that we use later on, i.e. essentially the Borel measures on X with values in E'_s.

DEFINITIONS III.2.1. A _Borel measure on_ X _with values in_ E is a function

$$m : \underline{B}(X) \to E.$$

It is

a) _finitely additive_ if we have $m(\emptyset) = 0$ and if, for every $b \in \underline{B}(X)$ and every finite Borel partition $\{b_1, \ldots, b_J\}$ of b, we have

$$m(b) = \sum_{j=1}^{J} m(b_j),$$

b) _countably additive_ if is finitely additive and if, for every $b \in \underline{B}(X)$ and every infinite countable Borel partition $\{b_j : j \in \mathbb{N}\}$ of b, we have

$$m(b) = \sum_{j=1}^{\infty} m(b_j),$$

the series converging in E.

PROPOSITION III.2.2. _If_ m _is a countably additive Borel measure on_ X _with values in_ E'_s, _then for every_ $e \in E$, _the function_

$$m_e : \underline{B}(X) \to \mathbb{C}$$

defined by

$$m_e(b) = \langle e, m(b) \rangle, \quad \forall\, b \in \underline{B}(X),$$

is a countably additive scalar Borel measure on X.

Proof. Of course we have

$$m_e(\emptyset) = \langle e, m(\emptyset)\rangle = 0$$

and, for every finite Borel partition $\{b_1, \ldots, b_J\}$ of $b \in \underline{B}(X)$,

$$\sum_{j=1}^{J} m_e(b_j) = \langle e, \sum_{j=1}^{J} m(b_j)\rangle = \langle e, m(b)\rangle = m_e(b).$$

Moreover let now $\{b_j : j \in \mathbb{N}\}$ be an infinite Borel partition of $b \in \underline{B}(X)$. For every $J \in \mathbb{N}$, we certainly have

$$\sum_{j=1}^{J} m_e(b_j) = \langle e, \sum_{j=1}^{J} m(b_j)\rangle.$$

Therefore as the series $\Sigma_{j=1}^{\infty} m(b_j)$ converges in E'_s, we get

$$\sum_{j=1}^{\infty} m_e(b_j) = \langle e, \sum_{j=1}^{\infty} m(b_j)\rangle = \langle e, m(b)\rangle = m_e(b).$$

Hence the conclusion.///

DEFINITIONS III.2.3. A countably additive Borel measure m on X with values in E'_s

a) is <u>regular</u> if for every $e \in E$, the scalar Borel measure m_e is regular,

b) <u>has a variation for</u> $p \in P$ (we also say "<u>has a</u> p-<u>variation</u>") <u>on</u> $b \in \underline{B}(X)$ if we have

$$\sup_{P \in \mathcal{P}(b)} \sum_{b' \in P} \|m(b')\|_p < \infty$$

where $\mathcal{P}(b)$ is the set of all the finite Borel partitions of b and where we have set

$$\|m(b')\|_p = \sup\{|\langle e, m(b')\rangle| : p(e) \leqslant 1\},$$

c) <u>has a variation for</u> $p \in P$ (we also say "<u>has a</u> p-<u>variation</u>") if it has a p-variation on every $b \in \underline{B}(X)$.

NOTATION III.2.4. If the countably additive Borel measure m on X with values in E'_s has a p-variation on $b \in \underline{B}(X)$, we set

$$V_p m(b) = \sup_{P \in P(b)} \sum_{b' \in P} \|m(b')\|_p$$

where $P(b)$ is the set of all the finite Borel partitions of b.

Therefore if the countably additive Borel measure m on X with values in E'_s has a p-variation, $V_p m$ appears as a scalar Borel measure on X with values in $[0,+\infty[$. This can be improved a lot.

PROPOSITION III.2.5. <u>If the countably additive Borel measure</u> m <u>on</u> X <u>with values in</u> E'_s <u>has a</u> p-<u>variation, then</u> $V_p m$ <u>is a countably additive scalar Borel measure on</u> X.

<u>Proof</u>. Of course $V_p m(\emptyset)$ is equal to 0.

We shall now establish that for every $b \in \underline{B}(X)$ and every infinite countable Borel partition $\{b_j : j \in \mathbb{N}\}$ of b, on one hand the series $\Sigma_{j=1}^{\infty} V_p m(b_j)$ converges and that its limit is less than or equal to $V_p m(b)$ and that on the other hand we also have

$$V_p m(b) \leq \sum_{j=1}^{\infty} V_p m(b_j)$$

hence the equality. As a simplification of this proof gives the equality for the finite Borel partitions of b, the proof will then be complete.

On one hand let us fix $\varepsilon > 0$ and remark that for every $j \in \mathbb{N}$, there is a finite Borel partition P_j of b_j such that

$$V_p m(b_j) \leq \sum_{b' \in P_k} \|m(b')\|_p + \varepsilon 2^{-j}.$$

As for every $J \in \mathbb{N}$ we certainly have

$$\sum_{j=1}^{J} \Big(\sum_{b' \in P_j} \|m(b')\|_p + \varepsilon 2^{-j} \Big) \leq V_p m(b) + \varepsilon$$

we get then

$$\sum_{j=1}^{\infty} V_p m(b_j) \leq V_p m(b) + \varepsilon.$$

Hence the conclusion of the first part.

To establish the second part it suffices to prove that for every finite Borel partition $\{b'_1, \ldots, b'_K\}$ of b we have

$$\sum_{k=1}^{K} \|m(b'_k)\|_p \leq \sum_{j=1}^{\infty} V_p m(b_j).$$

As m is countably additive and as for every $k \leq K$,

$$\{b'_k \cap b_j \neq \emptyset : j \in \mathbb{N}\}$$

is a countable Borel partition of b'_k, the series $\Sigma_{j=1}^{\infty} m(b'_k \cap b_j)$ converges in E'_s to $m(b'_k)$. This implies the inequality

$$\|m(b'_k)\|_p \leq \sum_{j=1}^{\infty} \|m(b'_k \cap b_j)\|_p, \quad \forall\, k \leq K.$$

Therefore we successively get

$$\sum_{k=1}^{K} \|m(b'_k)\|_p \leq \sum_{j=1}^{\infty} \sum_{k=1}^{K} \|m(b'_k \cap b_j)\|_p$$

$$\leq \sum_{j=1}^{\infty} V_p m(b_j). ///$$

THEOREM III.2.6. _Let_ m _be a countably additive Borel measure on_ X _with values in_ E'_s.

If there are $p \in P$, $C > 0$ _and_ a _compact subset_ K _of_ X _such that_ $m(b) \in C\, b_p^{\triangle}$ [$b_p^{\triangle}$ is the polar set of the semi-ball $b_p = \{e \in E : p(e) \leq 1\}$] _for every_ $b \in \underline{B}(X)$, $V_p m(X) < \infty$ _and_ $m(b) = 0$ _for every_ $b \in \underline{B}(X)$ _disjoint from_ K, _then_

a) m _has a_ p-_variation and_ $V_p m$ _is a countably additive scalar Borel measure on_ X _such that_ $V_p m(X) = V_p m(K)$,

b) _every_ $f \in C(X)$ _is_ m-_Riemann-integrable_; $\int f dm$ _belongs to_ E'_p _and is such that_

$$\|\int f dm\|_p \leq \int_X |f(x)|\, dV_p m \leq V_p m(K) . \|f\|_K,$$

c) _every_ $\varphi \in C(X;E)$ _is_ m-_Riemann-integrable_; $\int \varphi dm$ _is a complex number_

<u>such that</u>

$$|\int\varphi dm| \leq \int_X p(\varphi(x))dV_p m \leq V_p m(K).p_K(\varphi),$$

d) <u>for every</u> $f \in C(X)$ <u>and</u> $e \in E$, <u>we have</u>

$$\langle e,\int f dm\rangle = \int fe dm.$$

<u>Proof</u>. a) The measure m has a p-variation since every finite Borel partition of $b \in \underline{B}(X)$ is trivially included in a finite Borel partition of X and since we know that $V_p m(X)$ is finite. Therefore by the foregoing proposition, $V_p m$ is a countably additive scalar Borel measure on X. Finally we have $V_p m(K) = V_p m(X)$ because $X \setminus K$ is certainly a Borel subset of X such that $V_p m(X\setminus K) = 0$.

b) Let us remark first that for every $\varepsilon > 0$, there is a finite Borel partition P of K such that, for every $b \in P$ and every $x, y \in b$, we have $|f(x) - f(y)| \leq \varepsilon$. In fact, it is easy to get a finite open cover $\{\Omega_1, \ldots, \Omega_J\}$ of K such that, for every $j \leq J$ and every x, $y \in \Omega_j$, we have $|f(x) - f(y)| \leq \varepsilon$. It is then enough to take the non void subsets amongst

$$\Omega_1 \cap K,\ (\Omega_2 \setminus \Omega_1) \cap K,\ [\Omega_3 \setminus (\Omega_1 \cup \Omega_2)] \cap K,$$

$$\ldots,\ [\Omega_J \setminus (\Omega_1 \cup\ldots\cup \Omega_{J-1}) \cap K.$$

We are then naturally led into the following construction. Let ε_k be a sequence of positive numbers decreasing to 0. By the preceding remark, there is a sequence P_k of finite Borel partitions of K such that, for every $k \in \mathbb{N}$, $b \in P_k$ and $x, y \in b$, we have $|f(x)-f(y)| \leq \varepsilon_k$. By use of intersections, we may even suppose that the partitions P_k become finer as k increases. Finally for every $k \in \mathbb{N}$ and $b \in P_k$, we can choose a point $x_b \in b$.

i) We establish first that

$$\sum_{b\in P_k} f(x_b)m(b) \qquad (\star)$$

is a Cauchy sequence in the Banach space E'_p, i.e. we have

$$\sup_{p(e)=1} \left| \sum_{b\in P_r} f(x_b)\langle e,m(b)\rangle - \sum_{b'\in P_s} f(x_{b'})\langle e,m(b')\rangle \right| \to 0$$

if inf $\{r,s\} \to \infty$. Of course we may suppose to have $r \leqslant s$. As P_s is finer than P_r, for every $b \in P_r$, $\{b' \in P_s : b' \subset b\}$ is a finite Borel partition of b. Therefore we have

$$m(b) = \sum_{\substack{b' \in P_s \\ b' \subset b}} m(b'), \quad \forall\, b \in P_r.$$

This implies

$$\sup_{p(e)=1} \Big| \sum_{b \in P_r} f(x_b)\langle e, m(b)\rangle - \sum_{b' \in P_s} f(x_{b'})\langle e, m(b')\rangle \Big|$$

$$= \sup_{p(e)=1} \Big| \sum_{b \in P_r} \sum_{\substack{b' \in P_s \\ b' \subset b}} [\, f(x_b) - f(x_{b'})]\langle e, m(b')\rangle \Big|$$

$$\leqslant \sum_{b \in P_r} \sum_{\substack{b' \in P_s \\ b' \subset b}} |f(x_b) - f(x_{b'})| \sup_{p(e)=1} |\langle e, m(b')\rangle|$$

$$\leqslant \varepsilon_r \sum_{b' \in P_s} \| m(b') \|_p \leqslant \varepsilon_r \, V_p m(K).$$

Hence the conclusion.

ii) We establish now that the limit of the sequence (*) does not depend on the choice of the sequences ε_k and P_k and of the elements x_b. To prove this we consider, with notations clear by themselves, sequences ε'_k, ε''_k, P'_k and P''_k as well as points $x'_{b'} \in b'$ and $x''_{b''} \in b''$. For every $k \in \mathbb{N}$,

$$\{b' \cap b'' \neq \emptyset : b' \in P'_k,\ b'' \in P''_k\}$$

is a finite Borel partition of K. Therefore we successively get the following equality

$$\sup_{p(e)=1} \Big| \sum_{b' \in P'_k} f(x'_{b'})\langle e, m(b')\rangle - \sum_{b'' \in P''_k} f(x''_{b''})\langle e, m(b'')\rangle \Big|$$

$$= \sup_{p(e)=1} \Big| \sum_{\substack{b' \in P'_k \\ b'' \in P''_k}} [\, f(x'_{b'}) - f(x''_{b''})]\langle e, m(b' \cap b'')\rangle \Big|$$

hence the following inequalities

$$\sup_{p(e)=1} \left| \sum_{b'\in P'_k} f(x'_{b'})\langle e,m(b')\rangle - \sum_{b''\in P''_k} f(x''_{b''})\langle e,m(b'')\rangle \right|$$

$$\underset{(*)}{\leqslant} (\varepsilon'_k + \varepsilon''_k) \sum_{\substack{b'\in P'_k \\ b''\in P''_k}} \|m(b' \cap b'')\|_p$$

$$\leqslant (\varepsilon'_k + \varepsilon''_k)\, V_p m(K)$$

[in (*), we note that if $b' \cap b''$ is void, the corresponding term vanishes and that if $b' \cap b''$ is not void, it contains a point $x_{b'\cap b''}$ such that

$$|f(x_{b'}) - f(x_{b''})| \leqslant |f(x_{b'}) - f(x_{b',b''})| + |f(x_{b',b''}) - f(x_{b''})|$$

$$\leqslant \varepsilon'_k + \varepsilon''_k] .$$

Hence the conclusion.

So far we have proved that every $f \in C(X)$ is m-Riemann-integrable and that its m-integral $\int f dm$ belongs to E'_p and is the limit in that space E'_p of any suitable sequence

$$\sum_{b\in P_k} f(x_b)m(b).$$

The conclusion is then a direct consequence of the following relations

$$\|\int f dm\|_p = \lim_k \|\sum_{b\in P_k} f(x_b)m(b)\|_p$$

$$\leqslant \lim_k \sum_{b\in P_k} |f(x_b)|\, \|m(b)\|_p$$

$$\leqslant \|f\|_K\, V_p m(K).$$

c) As in the proof of b), one gets easily that for every $\varepsilon > 0$, there is a finite Borel partition P of K such that for every $b \in P$ and $x, y \in b$, one has $p(\varphi(x) - \varphi(y)) \leqslant \varepsilon$.

Now the construction goes as follows. Let ε_k be a sequence of positive numbers decreasing to 0. For every $k \in \mathbb{N}$, there is then a finite Borel partition P_k of K such that, for every $b \in P_k$ and $x, y \in b$, we have $p(\varphi(x) - \varphi(y)) \leq \varepsilon_k$. By use of intersections, we may even suppose that the partitions P_k become finer and finer. Finally for every $k \in \mathbb{N}$ and $b \in P_k$, we can choose a point $x_b \in b$. Just as in b), we can prove then that

$$\sum_{b \in P_k} \langle \varphi(x_b), m(b) \rangle$$

is a Cauchy sequence in $\mathbb{C}$ and that its limit does not depend on the choice of the sequences ε_k and P_k nor on the choice of the elements x_b.

This proves that every $\varphi \in C(X;E)$ is m-Riemann-integrable and that its m-integral $\int \varphi \, dm$ belongs to $\mathbb{C}$.

To conclude, let us remark that we certainly have the following relations

$$\begin{aligned} \left|\int \varphi \, dm\right| &= \lim_{k\to\infty} \left| \sum_{b \in P_k} \langle \varphi(x_b), m(b) \rangle \right| \\ &\leq \lim_{k\to\infty} \sum_{b \in P_k} p(\varphi(x_b)) \, \|m(b)\|_p \\ &\leq \lim_{k\to\infty} \sum_{b \in P_k} p(\varphi(x_b)) \, V_p m(b) \end{aligned}$$

for every $\varphi \in C(X;E)$ hence

$$\left|\int \varphi \, dm\right| \leq \int p(\varphi) \, dV_p m$$

by use of b), i.e. the interpretation of the $V_p m$-Riemann-integral of $p(\varphi) \in C(X)$.

Finally the inequality

$$\int p(\varphi) dV_p m \leq p_K(\varphi) V_p m(K)$$

is a direct consequence of b) applied to $p(\varphi)$.

d) is direct since, with notations clear by themselves, we have

$$\langle e, \int f dm \rangle = \lim_k \langle e, \sum_{b \in P_k} f(x_b) m(b) \rangle$$

hence

$$<e,\int f dm> = \lim_k \sum_{b\in P_k} <f(x_b)e,m(b)>$$

$$= \int fe dm. ///$$

PROPOSITION III.2.7. *Let* m_1, m_2 *be regular countably additive Borel measures on* X *with values in* E_s' *for which there are* $p \in P$, $C > 0$ *and a compact subset* K *of* X *such that for every* $j \in \{1,2\}$, *we have* $m_j(b) \in C\, b_p^\Delta$ *for every* $b \in \underline{B}(X)$, $V_p m_j(X) < \infty$ *and* $m_j(b) = 0$ *for every* $b \in \underline{B}(X)$ *disjoint from* K.

Then for every $e \in E$, *every open subset* Ω *of* X *and every* $b \in \underline{B}(X)$ *relatively compact in* Ω, *there is a sequence* f_k *of continuous functions on* X *with values in* $[0,1]$ *and with supports contained in* Ω *such that*

$$\int f_k e dm_1 \to <e,m_1(b)>$$

and

$$\int f_k e dm_2 \to <e,m_2(b)>.$$

Proof. Let us fix an element e of E, an open subset Ω of X and a Borel subset b of X such that $\bar{b}$ is a compact subset of Ω.

As the countably additive Borel measures $Vm_{1,e}$ and $Vm_{2,e}$ are regular, for every $k \in \mathbb{N}$, there are a closed subset and an open subset Ω_k of X such that

$$F_k \subset b \subset \Omega_k$$

and

$$V_p m_{j,e}(\Omega_k \setminus F_k) \leq \frac{1}{k}, \quad \forall\, j \in \{1,2\}.$$

Moreover substituting $\Omega_k \cap \Omega$ to Ω_k, we may suppose to have $\Omega_k \subset \Omega$ for every $k \in \mathbb{N}$. As it is contained in $\bar{b}$, each F_k is a compact subset of Ω_k. Now for every $k \in \mathbb{N}$, there is $f_k \in C(X)$ with values in $[0,1]$ which is identically 1 on a neighborhood of F_k and 0 on a neighborhood of $X \setminus \Omega_k$. In particular the support of f_k is always contained in Ω_k.

Therefore for $j \in \{1,2\}$ and with notations which are clear by themselves, we certainly have

$$|\int f_k e dm_j - \langle e, m_j(b)\rangle|$$

$$= \lim_l |\sum_{b' \in P_{j,l}} \{f_k(x_{b'})\langle e, m_j(b')\rangle - \langle e, m_j(b \cap b' \cap K)\rangle\}|.$$

Now let us remark that we may ask that each $b' \in P_{j,l}$ is contained in one of the sets

$$F_k \cap K,\ (b \setminus F_k) \cap K,\ (\Omega_k \setminus b) \cap K \text{ or } K \setminus \Omega_k.$$

Doing this provides the following inequalities

$$|\int f_k e dm_j - \langle e, m_j(b)\rangle|$$

$$= \lim_l \left| \sum_{\substack{b' \in P_{j,l} \\ b' \subset (b \setminus F_k) \cap K}} [f_k(x_{b'}) - 1]\langle e, m_j(b')\rangle + \sum_{\substack{b' \in P_{j,l} \\ b' \subset (\Omega_k \setminus b) \cap K}} f_k(x_{b'})\langle e, m_j(b')\rangle \right|$$

$$\leqslant \lim_l \sum_{\substack{b' \in P_{j,l} \\ b' \subset (\Omega_k \setminus F_k) \cap K}} |\langle e, m_j(b')\rangle|$$

$$\leqslant \lim_l \sum_{\substack{b' \in P_{j,l} \\ b' \subset (\Omega_k \setminus F_k) \cap K}} Vm_{j,e}(b')$$

$$\leqslant Vm_{j,e}(\Omega_k \setminus F_k) \leqslant \frac{1}{k}.$$

Hence the conclusion.///

III.3. STRUCTURE OF THE DUAL OF $C_c(X;E)$ OR THE SINGER THEOREM

PROPOSITION III.3.1. *For every* $\varphi \in C(X;E)$, *every* $p \in P$ *and every compact subset* K *of* X, *the function*

$$l_\varphi : K \times b_p^\Delta \to \mathbb{C}$$

defined by

$$l_\varphi(x,e') = \langle \varphi(x), e' \rangle, \quad \forall (x,e') \in K \times b_p^\Delta,$$

is continuous (b_p^Δ is considered as a topological subspace of E_s').

Proof. This is direct and well known. For every $x_o \in K$, $e_o' \in b_p^\Delta$ and $\varepsilon > 0$,

$$x \in K \quad \& \quad p(\varphi(x_o) - \varphi(x)) \leq \frac{\varepsilon}{2}$$

and

$$e' \in b_p^\Delta \quad \& \quad |\langle \varphi(x_o), e_o' - e' \rangle| \leq \frac{\varepsilon}{2}$$

imply

$$|l_\varphi(x_o,e_o') - l_\varphi(x,e')|$$

$$\leq |\langle \varphi(x_o), e_o' - e' \rangle| + |\langle \varphi(x_o) - \varphi(x), e' \rangle| \leq \varepsilon. ///$$

THEOREM III.3.2. *Let* φ' *be a linear functional on* $C(X;E)$. *Then* $p \in P$, $C > 0$ *and the compact subset* K *of* X *are such that*

$$|\langle \varphi, \varphi' \rangle| \leq C\, p_K(\varphi), \quad \forall \varphi \in C(X;E),$$

if and only if there is a regular countably additive scalar Borel measure $\mu_{\varphi'}$ *on the compact space* $K \times b_p^\Delta$ *such that*

$$\langle \varphi, \varphi' \rangle = \int l_\varphi \, d\mu_{\varphi'}, \quad \forall \varphi \in C(X;E),$$

where l_φ is defined as in the preceding proposition.

Proof. Let us remark first that

$$L = \{l_\varphi : \varphi \in C(X;E)\}$$

is a linear subspace of $C(K \times \overset{\Delta}{b}_p)$.

The condition is necessary. If φ, $\psi \in C(X;E)$ are such that $l_\varphi = l_\psi$ then we have $\langle\varphi(x),e'\rangle = \langle\psi(x),e'\rangle$ for every $x \in K$ and $e' \in \overset{\Delta}{b}_p$ hence $p_K(\varphi - \psi) = 0$ which implies $\langle\varphi,\varphi'\rangle = \langle\psi,\varphi'\rangle$. Therefore we can define a linear functional l' on L by

$$\langle l_\varphi, l'\rangle = \langle\varphi,\varphi'\rangle, \quad \forall\, \varphi \in C(X;E),$$

and it is immediate that l' verifies also

$$|\langle l,l'\rangle| \leqslant C \sup_{(x,e')\in K\times\overset{\Delta}{b}_p} |l(x,e')|, \quad \forall\, l \in L.$$

Hence the conclusion by use of the Hahn-Banach theorem to extend l' to $C(K\times\overset{\Delta}{b}_p)$ and of the Riesz theorem (let us recall that $K\times\overset{\Delta}{b}_p$ is compact).

The sufficiency of the condition is obvious.///

Remark III.3.3. In the proof of the necessity of the condition in the last theorem, we have used the Hahn-Banach theorem. Therefore the representation of the continuous linear functionals on $C_c(X;E)$ that we have got there, may very well not be unique. To this effect, the following result is more interesting. It will also enable us to introduce in the next chapter a notion of support for those functionals.

THE SINGER THEOREM III.3.4.([23], [29]) *Let φ' be a linear functional on* $C(X;E)$.

Then $p \in P$, $C > 0$ *and the compact subset* K *of* X *are such that*

$$|\langle\varphi,\varphi'\rangle| \leqslant C\, p_K(\varphi), \quad \forall\, \varphi \in C(X;E),$$

if and only if there is a regular countably additive Borel measure $m_{\varphi'}$ *on* X *with values in* E'_s *such that* $m_{\varphi'}(b) \in C\, \overset{\Delta}{b}_p$ *for every* $b \in \mathcal{B}(X)$, $V_p m_{\varphi'}(X) \leqslant C$, $m_{\varphi'}(b) = 0$ *for every* $b \in \mathcal{B}(X)$ *disjoint from* K *and*

$$\langle\varphi,\varphi'\rangle = \int \varphi\, dm_{\varphi'}, \quad \forall\, \varphi \in C(X;E).$$

Moreover this representation is unique.

Proof. The sufficiency of the condition is a direct consequence of part c) of the theorem III.2.6.

Let us now prove that the condition is necessary.

For every $e \in E$, we may define a linear functional φ'_e on $C(K)$ by

$$<f,\varphi'_e> = <\tilde{f}e,\varphi'>, \quad \forall f \in C(K),$$

where $\tilde{f}$ is any continuous extension of $f \in C(K)$ on X because on one hand as K is a compact subset of X, we know that there is at least one such extension and on the other hand, if $g, h \in C(X)$ are extensions of $f \in C(K)$, we have

$$|<ge - he,\varphi'>| \leqslant C\ p_K(ge - he) = 0.$$

Of course φ'_e is a linear functional on $C_c(K)$. Moreover it is continuous since for every $f \in C_c(K)$, we have

$$|<f,\varphi'_e>| = |<\tilde{f}e,\varphi'>| \leqslant C\ p_K(\tilde{f}e) = C\ p(e)\ \|f\|_K.$$

Therefore by use of the Riesz theorem, for every $e \in E$, there is a regular countably additive scalar Borel measure $\mu_{\varphi',e}$ on K such that

$$<fe,\varphi'> = <f|_K,\varphi'_e> = \int f|_K d\mu_{\varphi',e}, \quad \forall f \in C(X).$$

For every $b \in \underline{B}(X)$, $\mu_{\varphi',\cdot}(b \cap K)$ appears then as a linear functional on E that we shall rather write $m_{\varphi'}(b)$ so that we have

$$<e,m_{\varphi'}(b)> = \mu_{\varphi',e}(b \cap K), \quad \forall b \in \underline{B}(X), \forall e \in E.$$

At that time, $m_{\varphi'}$ appears as a Borel measure on X with values in $E^\star_s$ [i.e. the algebraic dual $E^\star$ of E endowed with the $\sigma(E^\star,E)$-topology] and it is direct matter to verify that it is finitely additive. In fact we have a lot more.

i) $m_{\varphi'}$ is countably additive.

As it is finitely additive, it is enough to check that for every countable Borel partition $\{b_k : k \in \mathbb{N}\}$ of a Borel subset b of X, we successively have

$$<e,m_{\varphi'}(b)> = \mu_{\varphi',e}(b \cap K)$$

$$= \sum_{k=1}^{\infty} \mu_{\varphi',e}(b_k \cap K)$$

$$= \sum_{k=1}^{\infty} <e,m_{\varphi'}(b_k)>, \quad \forall\, e \in E,$$

because this implies that the series $\Sigma_{k=1}^{\infty} m_{\varphi'}(b_k)$ converges in $E_s^{\star}$ to $m_{\varphi'}(b)$.

ii) $m_{\varphi'}$ <u>is regular</u>.

Let us fix $e \in E$. For every $b \in \underline{B}(X)$, we have

$$m_{\varphi',e}(b) = <e,m_{\varphi'}(b)> = \mu_{\varphi',e}(b \cap K).$$

As $\mu_{\varphi',e}$ is regular, for every $b \in \underline{B}(X)$ and $\varepsilon > 0$, there are a closed subset F of K and an open subset Ω' of K such that $F \subset b \cap K \subset \Omega'$ and $V\mu_{\varphi',e}(\Omega' \setminus F) \leqslant \varepsilon$. It is then direct to verify that the compact subset F of K and the open subset $\Omega = \Omega' \cup (X \setminus K)$ of X are such that $F \subset b \subset \Omega$ and $Vm_{\varphi',e}(\Omega \setminus F) \leqslant \varepsilon$. Hence the conclusion.

iii) <u>for every</u> $b \in \underline{B}(X)$, $m_{\varphi'}(b)$ <u>belongs to</u> $C\, b_p^{\Delta}$, i.e. we have $|<e,m_{\varphi'}(b)>| \leqslant C\, p(e)$ for every $e \in E$; in particular, $m_{\varphi'}$ takes its values in E_s'.

Let us fix $b \in \underline{B}(X)$ and $e \in E$. By use of ii), for every $\varepsilon > 0$, there are a compact subset K_ε of K and an open subset Ω_ε of X such that

$$K_\varepsilon \subset b \subset \Omega_\varepsilon \quad \text{and} \quad V\mu_{\varphi',e}[(\Omega_\varepsilon \setminus K_\varepsilon) \cap K] \leqslant \varepsilon.$$

There is then $f_\varepsilon \in C(X)$ with values in $[0,1]$ which is identically 1 on a neighborhood of K_ε and 0 on a neighborhood of $X \setminus \Omega_\varepsilon$. For such an f_ε, we have

$$\left|\int f_\varepsilon|_K \, d\mu_{\varphi',e} - \mu_{\varphi',e}(b \cap K)\right| \leqslant \left|\int (f_\varepsilon|_K - \chi_{b \cap K})\, d\mu_{\varphi',e}\right|$$

$$\leqslant V\mu_{\varphi',e}[(\Omega_\varepsilon \setminus K_\varepsilon) \cap K] \leqslant \varepsilon.$$

Therefore we get

$$|\langle e, m_{\varphi'}(b)\rangle| = |\mu_{\varphi',e}(b \cap K)|$$

$$\leqslant \sup_{\substack{f \in C(X) \\ \|f\|_K \leqslant 1}} \left|\int f|_K \, d\mu_{\varphi',e}\right|$$

$$\leqslant \sup_{\substack{f \in C(X) \\ \|f\|_K \leqslant 1}} |\langle fe, \varphi'\rangle| \leqslant C\ p(e).$$

iv) $V_p m_{\varphi'}(X) \leqslant C$.

To get this we just have to prove that we have

$$\sum_{j=1}^{J} \|m_{\varphi'}(b_j)\|_p \leqslant C$$

for every finite Borel partition $\{b_1, \ldots, b_J\}$ of X.

Let us fix $\varepsilon > 0$. For every $j \leqslant J$, there are on one hand [by use of the definition of $\|m_{\varphi'}(b_j)\|_p$] an element e_j of E such that

$$p(e_j) = 1 \quad \text{and} \quad \|m_{\varphi'}(b_j)\|_p \leqslant |\langle e_j, m_{\varphi'}(b_j)\rangle| + \frac{\varepsilon}{3J}$$

and on the other hand [by use of ii)] a compact subset K_j of K and an open subset Ω_j of X such that

$$K_j \subset b_j \subset \Omega_j \quad \text{and} \quad V\mu_{\varphi',e_j}[(\Omega_j \setminus K_j) \cap K] \leqslant \frac{\varepsilon}{3J}.$$

Now let us choose $f_j \in C(X)$ with values in [0,1] which is identically 1 on a neighborhood of K_j and 0 on a neighborhood of $X \setminus \Omega_j$. We get

$$\left|\mu_{\varphi',e_j}(K_j) - \int f_j|_K \, d\mu_{\varphi',e_j}\right| \leqslant V\mu_{\varphi',e_j}[(\Omega_j \setminus K_j) \cap K] \leqslant \frac{\varepsilon}{3J}.$$

Therefore we successively have

$$\sum_{j=1}^{J} \|m_{\varphi'}(b_j)\|_p \leqslant \sum_{j=1}^{J} |\langle e_j, m_{\varphi'}(b_j)\rangle| + \frac{\varepsilon}{3}$$

$$\leqslant \sum_{j=1}^{J} |\mu_{\varphi',e_j}(b_j \cap K)| + \frac{\varepsilon}{3}$$

hence

$$\sum_{j=1}^{J} \|m_{\varphi'}(b_j)\|_p$$

$$\leqslant \sum_{j=1}^{J} |\mu_{\varphi',e_j}(b_j \cap K) - \mu_{\varphi',e_j}(K_j)|$$

$$+ \sum_{j=1}^{J} |\mu_{\varphi',e_j}(K_j) - \int f_j|_K \, d\mu_{\varphi',e_j}|$$

$$+ \sum_{j=1}^{J} |\int f_j|_K \, d\mu_{\varphi',e_j}| + \frac{\varepsilon}{3}$$

$$\leqslant \sum_{j=1}^{J} |\int f_j|_K \, d\mu_{\varphi',e_j}| + \varepsilon$$

and by setting $c_j = \exp(-i \arg \int f_j|_K \, d\mu_{\varphi',e_j})$

$$\sum_{j=1}^{J} \|m_{\varphi'}(b_j)\|_p \leqslant \sum_{j=1}^{J} \int c_j f_j|_K \, d\mu_{\varphi',e_j} + \varepsilon$$

$$\leqslant |\langle \sum_{j=1}^{J} c_j f_j e_j, \varphi' \rangle| + \varepsilon$$

$$\leqslant C \, p_K(\sum_{j=1}^{J} c_j f_j e_j) + \varepsilon \underset{(\star)}{\leqslant} C + \varepsilon.$$

[($\star$) holds if we impose that the functions $f_1, \ldots, f_J$ have mutually disjoint supports, which is possible since the compact subsets $K_1, \ldots, K_J$ are mutually disjoint.] Hence the conclusion.

v) <u>for every</u> $b \in \mathcal{B}(X)$ <u>such that</u> $b \cap K = \emptyset$, $m_{\varphi'}(b)$ <u>is equal to</u> 0.

This is direct since for every $e \in E$, we have then

$$\langle e, m_{\varphi'}(b) \rangle = \mu_{\varphi',e}(b \cap K) = 0.$$

vi) <u>for every</u> $\varphi \in C(X;E)$, $\int \varphi \, dm_{\varphi'}$ <u>is equal to</u> $\langle \varphi, \varphi' \rangle$.

By hypothesis, φ' is a continuous linear functional on $C_c(X;E)$. By part c) of the theorem III.2.6, $\int . dm_{\varphi'}$ is also a continuous linear functional on $C_c(X;E)$. As by theorem I.5.3, we know that $C(X)\otimes E$ is dense in $C_c(X;E)$, these two continuous linear functionals are equal if and only if they coincide on $C(X)\otimes E$. Hence the conclusion since for every $f \in C(X)$ and $e \in E$, we have

$$\int fedm_{\varphi'} \underset{(\star)}{=} \int f|_K \, d\mu_{\varphi',e} = \langle fe,\varphi'\rangle$$

[to prove that the equality $(\star)$ holds, just use the Riemann interpretation of the integral].

To conclude, we still have to prove the uniqueness of this representation, i.e. if m is a regular countably additive Borel measure on X with values in E'_s such that $m(b) \in C\, b_p^{\Delta}$ for every $b \in \underline{B}(X)$, $V_p m(X) \leqslant C$, $m(b) = 0$ for every $b \in \underline{B}(X)$ disjoint from K and $\langle .,\varphi'\rangle = \int .dm$ on $C(X;E)$, then we must have $m = m_{\varphi'}$. As we have

$$\int fedm_{\varphi'} = \langle fe,\varphi'\rangle = \int fedm, \quad \forall\, f \in C(X),\ \forall\, e \in E,$$

the proposition III.2.7 gives

$$\langle e,m_{\varphi'}(b)\rangle = \langle e,m(b)\rangle, \quad \forall\, e \in E,$$

hence $m_{\varphi'}(b) = m(b)$ for every $b \in \underline{B}(X)$ relatively compact in X. Hence the conclusion since for every $b \in \underline{B}(X)$, we also have

$$m_{\varphi'}(b) = m_{\varphi'}(b\cap K) \text{ and } m(b) = m(b\cap K).///$$

NOTATION III.3.5. If φ' is a continuous linear functional on $C_c(X;E)$, there are $p \in P$, $C > 0$ and a compact subset K of X such that $|\langle .,\varphi'\rangle| \leqslant C\, p_K(.)$ on $C_c(X;E)$. Then we denote by

$$m_{\varphi'}$$

the unique representation of φ' as a Borel measure given by the Singer theorem. Conversely if m is a Borel measure verifying the hypothesis of the theorem III.2.6, $\int .dm$ is a continuous linear functional on $C_c(X;E)$ that we designate by

$$\varphi'_m.$$

III.4. SUPPORT OF A SUBSET OF $C_c(X;E)'$

THEOREM III.4.1. *For every* $\varphi' \in C_c(X;E)'$, *there is a smallest compact subset* K *of* X *for which there are* $p \in P$ *and* $C > 0$ *such that*

$$|\langle\varphi,\varphi'\rangle| \leq C\, p_K(\varphi), \quad \forall\, \varphi \in C_c(X;E).$$

Moreover $x \in X$ *belongs to this smallest compact subset if and only if, for every neighborhood* V *of* x, *there are* $f \in C(X)$ *with support contained in* V *and* $e \in E$ *such that* $\langle fe,\varphi'\rangle \neq 0$.

In particular, this smallest compact subset coincides with the support $K(D_{\varphi'})$ *of*

$$D_{\varphi'} = \{\varphi \in C_c(X;E) : |\langle\varphi,\varphi'\rangle| \leq 1\}.$$

Proof. As φ' is a continuous linear functional on $C_c(X;E)$, there are $p \in P$, $C > 0$ and a compact subset K of X such that

$$|\langle .,\varphi'\rangle| \leq C\, p_K(.) \quad \text{on} \quad C_c(X;E).$$

By the Singer theorem there is then a regular countably additive Borel measure $m_{\varphi'}$ on X with values in E'_s such that $m_{\varphi'}(b) \in C\, b_p^{\Delta}$ for every $b \in \underline{B}(X)$, $V_p m_{\varphi'}(X) \leq C$, $m_{\varphi'}(b) = 0$ for every $b \in \underline{B}(X)$ disjoint from K and $\int .\,dm_{\varphi'} = \langle .,\varphi'\rangle$ on $C_c(X;E)$. Let us suppose now that $p_1 \in P$, $C_1 > 0$ and the compact subset K_1 of X are also such that $|\langle .,\varphi'\rangle| \leq C_1\, p_{1,K_1}(.)$ on $C_c(X;E)$. By the uniqueness of the representation of φ' by a measure, we get then $m_{\varphi'}(b) = 0$ for every $b \in \underline{B}(X)$ disjoint from $K \cap K_1$. This proves that we also have

$$|\langle\varphi,\varphi'\rangle| \leq C\, p_{K\cap K_1}(\varphi), \quad \forall\, \varphi \in C_c(X;E).$$

It is then standard to prove the existence of such a smallest compact subset.

Let us designate by K this smallest compact subset. On one hand if $x \in X$ belongs to K, for every open neighborhood V of x, the compact subset $K\setminus V$ is not suitable. Therefore there is $\varphi \in C_c(X;E)$ such that

$$C\, p_{K\setminus V}(\varphi) < |\langle\varphi,\varphi'\rangle| = |\textstyle\int \varphi\, dm_{\varphi'}|.$$

By the Riemann interpretation of the integral, this implies the existence of $b \in \underline{B}(X)$ such that $b \subset K \cap V$ and $m_{\varphi'}(b) \neq 0$. Hence there is $e \in E$ such that $\langle e, m_{\varphi'}(b)\rangle \neq 0$. By the proposition III.2.7, there is then a sequence f_k in $C(X)$ such that $\operatorname{supp} f_k \subset V$ for every $k \in \mathbb{N}$ and

$$\langle f_k e, \varphi'\rangle = \int f_k e \, dm_{\varphi'} \rightarrow \langle e, m_{\varphi'}(b)\rangle \neq 0$$

hence the conclusion of this part. On the other hand, if $x \in X$ does not belong to K, there is an open neighborhood V of x which is disjoint from K. Therefore for every $\varphi \in C_c(X;E)$ with support contained in V we have

$$|\langle \varphi, \varphi'\rangle| \leqslant C\, p_K(\varphi) = 0$$

hence the conclusion.

Now let us consider the particular case.

On one hand, the inclusion $K \subset K(D_{\varphi'})$ is a direct consequence of the criterion II.1.5. On the other hand if x belongs to $K(D_{\varphi'})$, for every neighborhood V of x, there is $\varphi \in C(X;E) \setminus D_{\varphi'}$ such that $\varphi^{\beta}(\beta X \setminus V) = \{0\}$. By the Riemann interpretation of the integral we get then the existence of $e \in E$ and $b \in \underline{B}(X)$ such that $b \subset K \cap V$ and $\langle e, m_{\varphi'}(b)\rangle \neq 0$. By the proposition III.2.7 there is then $f \in C(X)$ with support contained in V such that

$$\langle fe, \varphi'\rangle = \int fe \, dm_{\varphi'} \neq 0.$$

Hence the conclusion.///

DEFINITION III.4.2. For every $\varphi' \in C_c(X;E)'$,

$$\operatorname{supp} \varphi'$$

denotes the smallest compact subset of X obtained in the last theorem. It is called the <u>support of</u> φ'. It is also the support of the absolutely convex subset

$$\{\varphi \in C_c(X;E) : |\langle \varphi, \varphi'\rangle| \leqslant 1\}.$$

For every subset B of $C_c(X;E)'$, we set

$$\operatorname{supp} B = \cup \{\operatorname{supp} \varphi' : \varphi' \in B\};$$

it is a subset of X.

PROPOSITION III.4.3. <u>For every bounded subset</u> B <u>of</u> $C_c(X;E)'_s$, supp B <u>is a bounding subset of</u> X.

<u>Proof</u>. If it is not the case, there is a bounded subset B of $C_c(X;E)'_s$, $f \in C(X)$ and a sequence x_l of supp B such that

$$|f(x_{l+1})| > |f(x_l)| + 1, \quad \forall\, l \in \mathbb{N}.$$

There is then $\varphi'_1 \in B$ such that $x_1 \in \operatorname{supp} \varphi'_1$. As supp φ' is a compact subset of X for every $\varphi' \in C_c(X;E)'$, we can obtain by recurrence a subsequence x_{k_l} of the sequence x_l and a sequence $\varphi'_l \in B$ such that

$$x_{k_l} \in \operatorname{supp} \varphi'_l \setminus \bigcup_{j=1}^{l-1} \operatorname{supp} \varphi'_j, \quad \forall\, l \geqslant 2.$$

From the preceding theorem, for every $l \in \mathbb{N}$, we get easily the existence of $b_l \in \underline{B}(X)$ contained in the neighborhood

$$\{x \in X : |f(x) - f(x_{k_l})| \leqslant \tfrac{1}{4}\} \setminus \bigcup_{j=1}^{l-1} \operatorname{supp} \varphi'_j$$

of x_{k_l} such that $m_{\varphi'_l}(b_l) \neq \emptyset$. We may even impose that b_l is contained in supp φ'_l. Finally let e_l be a sequence of E such that

$$\langle e_l, m_{\varphi'_l}(b_l)\rangle \neq 0, \quad \forall\, l \in \mathbb{N}.$$

By use of the proposition III.2.7, we can get finally a sequence g_l of continuous functions on X with values in $[0,1]$ and with support contained in

$$\{x \in X : |f(x) - f(x_{k_l})| < \tfrac{1}{2}\} \setminus \bigcup_{j=1}^{l-1} \operatorname{supp} \varphi'_j$$

as well as a sequence $c_l \in \mathbb{C}$ such that

$$c_l \langle e_l, \int g_l \, dm_{\varphi'_l}\rangle = c_l \int g_l e_l \, dm_{\varphi'_l}, \quad \forall\, l \geqslant 2,$$

hence such that

$$c_l\langle e_l,\int g_l\, dm_{\varphi_l'}\rangle \geq \sum_{j=1}^{l-1} |\langle c_j g_j e_j,\varphi_l'\rangle| + 1, \quad \forall\, l \geq 2.$$

As the sets supp g_l ($l\in\mathbb{N}$) are locally disjoint, the series $\Sigma_{j=1}^{\infty} c_j g_j e_j$ belongs to $C_c(X;E)$. Hence a contradiction because the sequence φ_l' is bounded in $C_c(X;E)_s'$ and is such that

$$|\langle \sum_{j=1}^{\infty} c_j g_j e_j,\varphi_l'\rangle| = |\int \sum_{j=1}^{\infty} c_j g_j e_j\, dm_{\varphi_l'}|$$

$$= |\int \sum_{j=1}^{l} c_j g_j e_j\, dm_{\varphi_l'}|$$

$$= |\langle \sum_{j=1}^{l} c_j g_j e_j,\varphi_l'\rangle| \geq l, \quad \forall\, l \geq 2.///$$

PROPOSITION III.4.4. <u>If</u> P <u>is contained in</u> $K(X)$ <u>and if</u> B <u>is a bounded subset of</u> $C_P(X;E)_b'$, <u>then</u>

a) <u>every</u> P-<u>finite sequence of open subsets of</u> X <u>is finite on</u> supp B,

b) <u>the set</u>

$$\{m_{\varphi'}(b) : \varphi' \in B, b \in \underline{B}(X)\}$$

<u>is a bounded subset of</u> E_b'.

<u>Proof</u>. a) If it is not the case, there is a P-finite sequence Ω_l of open subsets of X such that $\Omega_l \cap$ supp $B \neq \emptyset$ for every $l \in \mathbb{N}$. Therefore there is a sequence $\varphi_l' \in B$ such that $\Omega_l \cap$ supp $\varphi_l' \neq \emptyset$ for every $l \in \mathbb{N}$ hence a sequence $b_l \in \underline{B}(X)$ such that

$$\overline{b_l} \subset \Omega_l \cap \text{supp}\, \varphi_l' \quad \text{and} \quad m_{\varphi_l'}(b_l) \neq 0, \quad \forall\, l \in \mathbb{N}.$$

From this we get the existence of a sequence e_l of E such that

$$|\langle e_l, m_{\varphi_l'}(b_l)\rangle| \geq l + 1, \quad \forall\, l \in \mathbb{N},$$

and of a sequence $f_l \in C(X)$ such that

$$|<f_l e_l, \varphi'_l>| \geq 1, \text{ supp } f_l \subset \Omega_l \quad \text{and} \quad f_l(X) \subset [0,1]$$

for every $l \in \mathbb{N}$. It is then immediate that $\{f_l e_l : l \in \mathbb{N}\}$ is a bounded subset of $C(X;E)$. Hence a contradiction.

b) If it is not the case, there is a bounded sequence e_l of E and sequences $\varphi'_l \in B$ and $b_l \in \underline{B}(X)$ such that

$$|<e_l, m_{\varphi'_l}(b_l)>| \geq l + 1, \quad \forall\, l \in \mathbb{N}.$$

From this we get the existence of a sequence f_l of continuous functions on X with values in $[0,1]$ such that

$$|<f_l e_l, \varphi'_l>| \geq l, \quad \forall\, l \in \mathbb{N}.$$

Hence a contradiction since that sequence $f_l e_l$ is certainly bounded in $C_P(X;E)$.///

PROPOSITION III.4.5. If P is contained in $K(X)$, then a subset B of $C_P(X;E)'$ is equicontinuous if and only if the following two conditions are satisfied :

a) there is $B \in P$ such that supp $B \subset \bar{B}^X$,

b) there are $p \in P$ and $C > 0$ such that, for every $\varphi' \in B$, we have $V_p m_{\varphi'}(X) \leq C$.

Proof. This is a direct consequence of the theorem III.2.6 and of the Singer theorem III.3.4.///

CHAPTER IV

CHARACTERIZATION OF LOCALLY CONVEX PROPERTIES OF THE $C_P(X;E)$ SPACES

IV.1. REFERENCE EXAMPLES

DEFINITIONS IV.1.1. Let us recall that X

a) is a k-*space* if the closed subsets of X coincide with the subsets of X which have compact intersection with every compact subset of X. Of course if X is a k-space, then a function on X with values in E is continuous as soon as it is continuous on every compact subset of X. In particular if X is a k-space and if E is complete [resp. quasi-complete; sq-complete], then $C_c(X;E)$ is complete [resp. quasi-complete; sq-complete].

b) is σ-*compact* if it is a countable union of compact subsets.

c) is *hemi-compact* if it has a countable fundamental family of compact subsets. In particular if X is hemi-compact, it is σ-compact. Conversely if X is locally compact and σ-compact, then it is hemi-compact.

THEOREM IV.1.2.

a) *The space* $C_P(X;E)$ *is normed if and only if* E *is normed and one element of* P *is dense in* υX (which implies that υX is compact).

In particular the space $C_c(X;E)$ *is normed if and only if* E *is normed and* X *is compact*.

b) *The space* $C_P(X;E)$ *is a Banach space if and only if* E *is a Banach space and one element of* P *is dense in* υX (which implies that υX is compact).

In particular the space $C_c(X;E)$ *is a Banach space if and only if* E *is a Banach space and* X *is compact*.

Proof. a) is a direct consequence of part b) of the proposition I.2.1.

b) is then immediate.///

THEOREM IV.1.3.

a) *The space* $C_{\mathcal{P}}(X;E)$ *is metrizable if and only if* E *is metrizable and* $\mathcal{P}$ *has a countable fundamental sequence* (in the following sense : there is a sequence $B_m \in \mathcal{P}$ such that for every $B \in \mathcal{P}$ there is $m \in \mathbb{N}$ such that $B \subset \overline{B_m}^{\upsilon X}$).

In particular the space $C_c(X;E)$ *is metrizable if and only if* E *is metrizable and* X *is hemi-compact*.

b) *If* X *is a* k-*space*, $C_c(X;E)$ *is a Fréchet space if and only if* E *is a Fréchet space and* X *is hemi-compact*.

Proof. a) is a direct consequence of part b) of the proposition I.2.1.

b) is then immediate.///

IV.2 $C_{\mathcal{P}}(X;E)$ AND $C_{Q,\mathcal{P}}(X;E)$ SPACES WHICH ARE MACKEY, ULTRABORNOLOGICAL, ... (NECESSARY CONDITION)

Remark IV.2.1. Let us recall briefly the following result : if L is a topologically complemented linear subspace of E and if E is a Mackey [resp. ultrabornological; bornological; barrelled; quasi-barrelled; (DF)] space, then L is a Mackey [resp. ultrabornological; bornological; barrelled; quasi-barrelled; (DF)] space.

The proof is direct : let P be a continuous linear projection from E onto L.

Let us consider the case where E is a Mackey space. The adjoint $P^* : L'_s \to E'_s$ is continuous and linear. Therefore, for every absolutely convex compact subset K of L'_s, P^*K is an absolutely convex and compact subset of E'_s hence is equicontinuous : there are $p \in P$ and $C > 0$ such that

$$\sup_{e' \in K} |\langle P.,e'\rangle| \leqslant C\, p(.)$$

on E. Hence the conclusion since we have $P|_L = Id_L$.

Let us now consider the case where E is ultrabornological [resp. bornological; barrelled; quasi-barrelled]. It is sufficient to prove that a semi-norm q on L is continuous if it is bounded on every absolutely convex compact subset of L [resp. bounded on every bounded subset of L; lower semi-continuous; lower semi-continuous and bounded on every bounded subset of L]. Then $q \circ P$ is a semi-norm on E which coincides with q on L and which is continuous on E since it is bounded on every absolutely convex compact subset of E [resp. bounded on every bounded subset of E; lower semi-continuous; lower semi-continuous and bounded on every bounded subset of E].

Now let E be a (DF) space. On one hand if B_m is a fundamental sequence of bounded subsets of E, then $B_m \cap L$ is of course a fundamental sequence of bounded subsets of L. On the other hand let q_m be a sequence of continuous semi-norms on L such that $q = \sup_m q_m$ is a semi-norm on L which is bounded on every bounded subset of L. Then $q_m \circ P$ is a sequence of continuous semi-norms on E such that $q \circ P = \sup_m q_m \circ P$ is a semi-norm on E which is bounded on every bounded subset of E. Hence $q \circ P$ is a continuous semi-norm on E which coincides with q on L.

THEOREM IV. 2.2. *If* $C_P(X;E)$ *is a Mackey* [resp. *ultrabornological; bornological; barrelled; quasi-barrelled;* (DF)] *space*

a) *then* $C_P(X)$ *is a Mackey* [resp. *ultrabornological; bornological; barrelled; quasi-barrelled;* (DF)] *space,*

b) *and if* $\cup \{\bar{B}^{\upsilon X} : B \in P\}$ *contains one point of* X, *then* E *is a Mackey* [resp. *ultrabornological; bornological; barrelled; quasi-barrelled;* (DF)] *space*.

Proof. This is a direct consequence of the theorem I.4.4 and of the previous remark.///

Just in the same way, we also get the following result.

THEOREM IV.2.2'. *If* $C_{Q,P}(X;E)$ *is a Mackey* [resp. *ultrabornological; bornological; barrelled; quasi-barrelled;* (DF)] *space, then* $C_P(X)$ *and* (E,Q) *are such spaces too*.///

The following very interesting examples show clearly that the foregoing two theorems have no general converse properties except for the (DF) property (cf. theorem IV.9.2).

THE S.-DIEROLF EXAMPLE IV.2.3 [26] _If_ K _is an infinite compact space and if_ E _is an uncountable dimensional linear space endowed with its finest locally convex topology, then_ $C_c(K;E)$ _is not quasi-barrelled_.

Proof. Under these conditions, we know that there is an uncountable set J such that E is topologically isomorphic to $\oplus_{j\in J} \mathbb{K}$ where $\mathbb{K}$ is $\mathbb{R}$ or $\mathbb{C}$. For every $k \in J$, let π_k be the k-th canonical projection from $\oplus_{j\in J} \mathbb{K}$ onto $\mathbb{K}$.

To conclude it is sufficient to establish that

$$T = \bigcap_{m\in\mathbb{N}} \bigcap_{\substack{j_1,\ldots,j_m\in J \\ |\{j_1,\ldots,j_m\}|=m \\ x_1,\ldots,x_m\in K}} \{\varphi \in C_c(K;E) : |\sum_{l=1}^{m} \pi_{j_l}[\varphi(x_l)]| \leq 1\}$$

is a bornivorous barrel of $C_c(K;E)$ which is not a neighborhood of 0 in that space.

a) T is an absolutely convex and closed subset of $C_c(K;E)$.

This is direct since for every $m \in \mathbb{N}$, every $j_1, \ldots, j_m \in J$ such that $|\{j_1, \ldots, j_m\}| = m$ and every $x_1, \ldots, x_m \in K$,

$$\sum_{l=1}^{m} \pi_{j_l}[\,.\,(x_l)]$$

is a continuous linear functional on $C_c(K;E)$.

b) T is bornivorous in $C_c(K;E)$.

Let B be a bounded subset of $C_c(K;E)$. Then

$$\{\varphi(x) : \varphi \in B,\ x \in K\}$$

is a bounded subset of E. Therefore there is a finite subset J' of J and a bounded subset B of $\mathbb{K}$ such that

$$\pi_j[\varphi(x)] = 0, \quad \forall\, x \in K,\ \forall\, j \in J\setminus J',$$

and

$$\pi_j[\varphi(x)] \in B, \quad \forall\, x \in K,\ \forall\, j \in J'\ .$$

Then for every $\varphi \in B$, every $m \in \mathbb{N}$, every $j_1, \ldots, j_m \in J$ such that $|\{j_1, \ldots, j_m\}| = m$ and every $x_1, \ldots, x_m \in K$, we have

$$\left|\sum_{l=1}^{m} \pi_{j_l}[\varphi(x_l)]\right| = \left|\sum_{\substack{1 \leq l \leq m \\ j_l \in J'}} \pi_{j_l}[\varphi(x_l)]\right|$$

$$\leq |J'| \sup_{c \in B} |c|$$

where the last member is finite; let us call it M. This proves that we have $\frac{1}{1+M} B \subset T$, hence the conclusion.

c) T contains no neighborhood of 0 in $C_c(K;E)$.

Let us consider the following neighborhood

$$b = \{\varphi \in C_c(K;E) : p_K(\varphi) \leq r\} \quad (p \in P; r > 0)$$

of 0 in $C_c(K;E)$. Without loss of generality we may of course suppose that p can be written $\Sigma_{j\in J}\, r_j |\pi_j \cdot|$ where all the r_j are positive numbers. As J is an uncountable set, there are then $\eta > 0$ and an infinite subset J' of J such that $r/r_j \geq \eta$ for every $j \in J'$. Now choose $m \in \mathbb{N}$ such $m\eta > 1$, distinct elements $j_1, \ldots, j_m$ of J' and distinct points $x_1, \ldots, x_m$ of K. Under these circumstances there are also continuous functions $f_1, \ldots, f_m$ on K with values in $[0,1]$, with mutually disjoint supports and such that $f_l(x_l) = 1$ for every $l \leq m$. Therefore

$$\sum_{k=1}^{m} \frac{r}{r_k} f_k\, \varepsilon_{j_k}$$

(where for every $k \in J$, ε_k is the k-th unit vector of $\oplus_{j\in J} \mathbb{K}$) belongs to $C_c(K;E)$ and even to b but does not belong to T since we have

$$\left|\sum_{l=1}^{m} \pi_{j_l}\left[\left(\sum_{k=1}^{m} \frac{r}{r_k} f_k\, \varepsilon_{j_k}\right)(x_l)\right]\right| = \left|\sum_{l=1}^{m} \frac{r}{r_l}\right| \geq m\eta > 1.$$

Hence the conclusion.///

THE MARQUINA-SANZ SERNA EXAMPLE IV.2.4. [12] <u>If</u> E <u>is an uncountable dimensional linear space endowed with its finest locally convex topology, then</u> $c_o(E)$ <u>is not quasi-barrelled</u>.

Proof. Let us denote by $K = \mathbb{N} \cup \{\infty\}$ the Alexandrof compactification of the discrete space $\mathbb{N}$. It is direct matter to verify that the map

$$T : C_c(K;E) \to c_o(E) \times E$$

defined by

$$T\varphi = [(\varphi(m) - \varphi(\infty))_{m \in \mathbb{N}}, \varphi(\infty)], \quad \forall \varphi \in C_c(K;E),$$

is an isomorphism. Hence the conclusion since we know that $C_c(K;E)$ is not quasi-barrelled.///

Remark IV.2.5. We refer to the remark IV.6.5 for another proof of this last example.

IV.3. MACKEY $C_{Q,s}(X;E)$ SPACES

PROPOSITION IV.3.1. The space $C_{Q,s}(X;E)$ is a Mackey space if and only if (E,Q) is a Mackey space.

In particular, $C_s(X;E)$ is a Mackey space if and only if E is a Mackey space.

Proof. The necessity of the condition has alreday been established in theorem IV.2.2'.

The condition is sufficient. Let $\mathcal{K}$ be an absolutely convex compact subset of $C_{Q,s}(X;E)'_s$. As $\mathcal{K}$ is a bounded subset of $C_{Q,s}(X;E)'_b$, we know already by the proposition II.7.5 that supp $\mathcal{K}$ is a finite set. Moreover for every $x \in$ supp $\mathcal{K}$, the map

$$T_x : C_{Q,s}(X;E)'_s \to (E,Q)'_s$$

defined by

$$T_x\varphi' = e'_{\varphi',x}, \quad \forall \varphi' \in C_{Q,s}(X;E)'_s,$$

is certainly linear. It is also continuous since for every $x \in$ supp $\mathcal{K}$, there is $f_x \in C(X)$ such that $f_x(y) = \delta_{x,y}$ for every $y \in \operatorname{supp} K$ hence such that

$$|\langle e, e'_{\varphi',x} \rangle| = |\langle f_x e, \varphi' \rangle|, \quad \forall\, e \in E.$$

Therefore for every $x \in \operatorname{supp} K$, $T_x K$ is an absolutely convex compact subset of $(E,Q)'_s$ hence is equicontinuous. Hence the conclusion by the proposition II.7.6.///

<u>Remark</u> IV.3.2. Let us recall that E <u>has the strong Mackey property</u> if for every bounded subset B of E, there is an absolutely convex, bounded and closed subset D of E containing B and such that E and E_D endow the same topology on B.

a) In [20;58], S. Warner has proved that $C_c(X)$ has the strong Mackey property if and only if X is hemi-compact.

b) In [14], J. Mendoza has proved that $C_c(X;E)$ has the strict Mackey property if and only if $C_c(X)$ and E have the strict Mackey property.

IV.4. BORNOLOGICAL $C_P(X;E)$ SPACES (SUFFICIENT CONDITION)

The following result reduces very much the search of the bornological $C_P(X;E)$ spaces to the one of the bornological $C_{K(\cup X)}(X;E)$ spaces and motivates the whole paragraph.

THEOREM IV.4.1. <u>If the space</u> $C_{K(\cup X)}(X;E)$ <u>is bornological, then one has</u>

$$C_{\tilde{P}^{\cup}}(X;E) = C_P(X;E)_b.$$

<u>In particular, if the space</u> $C_{K(\cup X)}(X;E)$ <u>is bornological, then</u> $C_P(X;E)$ <u>is bornological if and only if</u> $C_P(X)$ <u>is bornological</u>.

<u>Proof</u>. As P is contained in $\tilde{P}^{\cup}$, $C_{\tilde{P}^{\cup}}(X;E)$ has a finer topology than $C_P(X;E)$. However by the very definition of $\tilde{P}^{\cup}$, the spaces $C_P(X;E)$ and $C_{\tilde{P}^{\cup}}(X;E)$ have the same bounded subsets. Therefore to conclude, it

is sufficient to prove that $C_{\mathcal{P}^{\upsilon}}(X;E)$ is bornological.

Let D be an absolutely convex and bornivorous subset of $C_{\mathcal{P}^{\upsilon}}(X;E)$. As $\mathcal{P}^{\upsilon}$ is contained in $K(\upsilon X)$, D is also an absolutely convex and bornivorous subset of the bornological space $C_{K(\upsilon X)}(X;E)$ hence is a neighborhood of 0 in that space : there are $p \in P$, $r > 0$ and a compact subset K of υX such that

$$D \supset \{\varphi \in C(X;E) : p_K(\varphi) \leqslant r\}.$$

As by proposition II.5.2, K(D) belongs to $\mathcal{P}^{\upsilon}$, the theorem II.1.4 gives

$$D \supset \{\varphi \in C(X;E) : p_{K(D)}(\varphi) < r\},$$

i.e. D is a neighborhood of 0 in $C_{\mathcal{P}^{\upsilon}}(X;E)$. Hence the conclusion.///

THEOREM IV.4.2. _If_ E _is metrizable, then_ $C_{K(\upsilon X)}(X;E)$ _is bornological_.

In particular, if E _is metrizable_

a) $C_{\mathcal{P}^{\upsilon}}(X;E)$ _is the bornological space associated to_ $C_{P}(X;E)$,

b) $C_{P}(X;E)$ _is bornological if and only if_ $C_{P}(X)$ _is bornological_,

c) $C_c(X;E)$ _is bornological if and only if_ X _is realcompact_.

Proof. This is a direct consequence of theorem II.4.3.

The particular cases are then straightforward consequences of the previous result.///

The following result due to Mendoza is based on a sharp use of the results of chapter two.

THE MENDOZA THEOREM IV.4.3 [13] _If_ X _is realcompact and locally compact and if_ $C_c(\beta X;E)$ _is bornological, then_ $C_c(X;E)$ _is bornological_.

In particular if X _is realcompact and locally compact and if_ $C_c(\beta X;E)$ _is bornological, then_

a) $C_{\mathcal{P}^{\upsilon}}(X;E)$ _is the bornological space associated to_ $C_{P}(X;E)$,

b) $C_{P}(X;E)$ _is bornological if and only if_ $C_{P}(X)$ _is bornological_.

Proof. Let D be an absolutely convex and bornivorous subset of $C_c(X;E)$.

On one hand as $D \cap CRC_u(X;E)$ is an absolutely convex and bornivorous subset of $CRC_u(X;E)$ and as $CRC_u(X;E)$ is topologically isomorphic to the bornological space $C_c(\beta X;E)$ (cf. theorem I.8.8), there is $p \in P$ such that

$$\{\varphi \in CRC_u(X;E) : \sup_{x \in X} p(\varphi(x)) \leqslant 1\} \subset D \cap CRC_u(X;E).$$

On the other hand the corollary II.2.5 implies that K(D) is a compact subset of $\upsilon X = X$. To conclude it is then enough to prove that we have

$$\{\varphi \in C_c(X;E) : p_{K(D)}(\varphi) < 1\} \subset D.$$

Let $\varphi \in C_c(X;E)$ be such that $p_{K(D)}(\varphi) < 1$.

The open subset of X

$$\Omega = \{x \in X : p(\varphi(x)) < 1\}$$

contains the compact subset K(D) of X. As X is locally compact there is then an open neighborhood V of K(D) with compact closure contained in Ω. There is finally a continuous function f on X with values in $[0,1]$ which is identically 1 on a neighborhood of K(D) and 0 on a neighborhood of $X \setminus V$, hence having a compact support contained in Ω.

Now $f\varphi$ belongs of course to $C_c(X;E)$ and even to $CRC_u(X;E)$ since its support is contained in the compact subset supp f. Moreover

$$\sup_{x \in X} p(f(x)\varphi(x))$$

certainly belongs to $[0,1[$ hence is less than or equal to some $r \in]0,1[$ for which we have $\frac{1}{r} f\varphi \in D$. Let us also remark that $\frac{1}{1-r}(1-f)\varphi$ belongs to $C_c(X;E)$ and is identically 0 on a neighborhood of K(D) hence belongs to D. Hence the conclusion since we have then

$$\varphi = r \, \frac{1}{r} f\varphi + (1-r) \, \frac{1}{1-r} \, (1-f)\varphi \in D.$$

The particular cases are straightforward consequences of the theorem IV.4.1 because X is realcompact.///

The Mendoza theorem can be generalized in the following way by use of the Mujica theorem I.7.2.

THEOREM IV.4.4. *If* X *is realcompact and locally compact and if* Ind E_m (cf. our convention I.7.1) *is regularly compact and such that* $C_c(\beta X;E_m)$ *is bornological for every* $m \in \mathbb{N}$, *then*

a) $C_c(X;\text{Ind } E_m)$ *is bornological,*

b) $C_{\tilde{P}\upsilon}(X;\text{Ind } E_m)$ *is the bornological space associated to* $C_P(X;\text{Ind } E_m)$,

c) $C_P(X;\text{Ind } E_m)$ *is bornological if and only if* $C_P(X)$ *is bornological.*

Proof. By the corollary I.7.3, we know that the equality

$$\text{Ind } C_c(\beta X;E_m) = C_c(\beta X;\text{Ind } E_m)$$

holds algebraically and topologically. Therefore the space $C_c(\beta X;\text{Ind } E_m)$ is bornological as inductive limit of bornological spaces.

The conclusion is then a direct consequence of the previous Mendoza theorem.///

PROPOSITION IV.4.5. *If* X *is realcompact, if every compact subset of* X *is finite, if* (E,Q) *is bornological and if there is a countable system of semi-norms on* E *coarser than* P, *then* $C_{Q,c}(X;E)$ *is bornological.*

Proof. By proposition IV.3.1, we know already that $C_{Q,c}(X;E) = C_{Q,s}(X;E)$ is a Mackey space. To conclude it is then sufficient to prove that a linear functional φ' on $C_{Q,c}(X;E)$ is continuous if it is bounded on every bounded subset. By proposition II.6.3, supp φ' is a compact subset of υX hence a finite subset of X. Moreover by proposition II.4.5 we have $\langle \varphi,\varphi' \rangle = 0$ for every $\varphi \in C(X;E)$ such that $\varphi(\text{supp } \varphi') = 0$. Therefore we can define a linear functional e' on the bornological space $E^{\text{supp } \varphi'}$ by

$$\langle e,e' \rangle = \langle \varphi_e,\varphi' \rangle, \quad \forall\, e \in E^{\text{supp } \varphi'},$$

where φ_e is any element of $C(X;E)$ such that $\varphi_e(x) = e(x)$ for every $x \in \text{supp } \varphi'$. Of course e' is a linear functional on $E^{\text{supp } \varphi'}$; moreover it is easy to check that it is bounded on the bounded subsets of

$(E,Q)^{\text{supp }\varphi'}$. Hence e' is a continuous linear functional. The conclusion is then immediate.///

THEOREM IV.4.6. *If* $C_s(X)$ *and* (E,Q) *are bornological and if* X *satisfies the first axiom of countability, then* $C_{Q,s}(X;E)$ *is bornological. Moreover if* $C_s(X)$ *is bornological and if* X *satisfies the first axiom of countability, then the bornological space associated to* $C_{Q,s}(X;E)$ *is the space* $C_{Q_b,s}(X;E)$ *where* Q_b *is the system of the continuous semi-norms of* $(E,Q)_b$.

In particular if $C_s(X)$ *and* E *are bornological and if* X *satisfies the first axiom of countability, then* $C_s(X;E)$ *is bornological. Moreover if* $C_s(X)$ *is bornological and if* X *satisfies the first axiom of countability, then the bornological space associated to* $C_s(X;E)$ *is the space* $C_{P_b,s}(X;E)$ *where* P_b *is the system of the continuous semi-norms of* E_b.

Proof. By proposition IV.3.1, we know already that $C_{Q,s}(X;E)$ is a Mackey space. To conclude it is then sufficient to prove that a linear functional φ' on $C_{Q,s}(X;E)$ is continuous if it is bounded on every bounded subset. As $C_s(X)$ is bornological, we know by the corollary III.4.3 of [20] that X is realcompact. Therefore by the remark II.6.2.c), supp φ' is a finite subset of X such that $\langle\varphi,\varphi'\rangle = 0$ for every $\varphi \in C(X;E)$ such that $\varphi(\text{supp }\varphi') = 0$. One can then end the proof as in the proof of the previous proposition.

The characterization of the bornological space associated to $C_{Q,s}(X;E)$ is then a direct consequence of the following three facts :

a) we have just proved that $C_{Q_b,s}(X;E)$ is bornological,

b) $C_{Q_b,s}(X;E)$ certainly has a finer topology than $C_{Q,s}(X;E)$,

c) it is easy to check that $C_{Q,s}(X;E)$ and $C_{Q_b,s}(X;E)$ have the same bounded subsets.

The particular case is immediate.///

To be complete we mention without proof the following fine result that Defant and Govaerts have obtained by use of tensor product methods.

THEOREM IV.4.7 [3] *Let* K *be an infinite compact Hausdorff space and* $(F_\alpha,\pi_{\alpha\beta})_{\alpha\in A}$ *be an inductive net such that the following two conditions hold* :

a) $C_c(K;F_\alpha)$ *is bornological for every* $\alpha\in A$,

b) *for every* $\varphi \in C(K;\text{Ind } F_\alpha)$, *there are* $\alpha\in A$ *and* $\psi \in C(K,F_\alpha)$ *such that* $\varphi = \pi_\alpha \circ \psi$.

Then $C_c(K;\text{Ind } F_\alpha)$ *is bornological if and only if the strong dual of* Ind F_α *has property* (B).///

IV.5. ULTRABORNOLOGICAL $C_P(X;E)$ SPACES (SUFFICIENT CONDITION)

THEOREM IV.5.1. *If* X *is realcompact, if* $\cup \{B : B\in P\}$ *is equal to* X *and if* E *is a Fréchet space, then* $C_c(X;E)$ *is the ultrabornological space associated to* $C_P(X;E)$.

In particular if X *is realcompact and if* E *is a Fréchet space, then* $C_c(X;E)$ *is ultrabornological*.

Proof. The particular case is a direct consequence of the theorem II.4.3.

To conclude it is then enough to establish that every absolutely convex, bounded and completing subset B of $C_P(X;E)$ is bounded in $C_c(X;E)$ because $C_c(X;E)$ has of course a finer topology than $C_P(X;E)$. Let us remark that for every compact subset K of X, the set $B|_K$ of the restrictions to K of the elements of B is certainly bounded and absolutely convex in $C_s(K;E)$. Moreover it is completing : as B is completing, there is $C > 1$ such that for every sequence φ_m of B, the series $\Sigma_{m=1}^{\infty} 2^{-m} \varphi_m$ converges in $C_P(X;E)$ to some element of CB; this implies of course that for every sequence ψ_m of $B|_K$, the series $\Sigma_{m=1}^{\infty} 2^{-m} \psi_m$ converges in $C_s(K;E)$ to some element of $CB|_K$. As $C_c(K;E)$ is a Fréchet space which has a finer topology than $C_s(K;E)$, it is the ultrabornological space associated to $C_s(K;E)$ (cf. proposition I.3.3. of [20]). Therefore $C_s(K;E)$ and $C_c(K;E)$ have the same absolutely convex, bounded and completing subsets. Hence the conclusion.///

THE MENDOZA THEOREM IV.5.2 [13] *If* X *is realcompact and locally compact and if* $C_c(\beta X;E)$ *is ultrabornological, then* $C_c(X;E)$ *is ultrabornological*.

Proof. It is enough to change "absolutely convex and bornivorous subset" by "absolutely convex subset which absorbs every absolutely convex and compact subset" in the proof of the theorem IV.4.3.///

THEOREM IV.5.3. _If_ X _is realcompact and locally compact and if_ Ind E_m (cf. our convention I.7.1) _is regularly compact and such that_ $C_c(\beta X;E_m)$ _is ultrabornological for every_ $m \in \mathbb{N}$, _then_ $C_c(X;\text{Ind } E_m)$ _is ultrabornological_.

Proof. It is enough to proceed as in the proof of the theorem IV.4.4.///

THEOREM IV.5.4. _If_ X _is realcompact, if every compact subset of_ X _is finite and if_ E _is metrizable and ultrabornological, then_ $C_c(X;E)$ _is ultrabornological_.

Proof. By proposition IV.3.1, we know already that $C_c(X;E) = C_s(X;E)$ is a Mackey space. To conclude it is then sufficient to prove that a linear functional φ' on $C_c(X;E)$ is continuous if it is bounded on every absolutely convex compact subset. By part e) of the remarks II.6.2, supp φ' is a compact subset of υX, hence a finite subset of X, such that we have $\langle\varphi,\varphi'\rangle = 0$ for every $\varphi \in C(X;E)$ such that $\varphi(\text{supp } \varphi') = 0$. One can then end the proof as in the proof of the proposition IV.4.5.///

IV.6. QUASI-BARRELLED $C_P(X;E)$ SPACES (SUFFICIENT CONDITION)

THE MENDOZA THEOREM IV.6.1 [13] _If_ P _is contained in_ $K(X)$ _and if_ $C_P(X)$ _and_ $C_c(\beta X;E)$ _are quasi-barrelled, then_ $C_P(X;E)$ _is quasi-barrelled_.

Proof. Let T be a bornivorous barrel of $C_P(X;E)$.

The polar set T^Δ is then a bounded subset of $C_P(X;E)'_b$ which implies by the proposition III.4.3 that supp T^Δ is a bounding subset of X hence a relatively compact subset of υX. Moreover supp T^Δ is contained in $\cup \{\bar{B}^{\upsilon X} : B \in P\}$ since for every $\varphi' \in T^\Delta$, supp φ' is contained in one of the sets $\bar{B}^{\upsilon X}$ $(B \in P)$. Therefore by the proposition III.4.4, supp T^Δ belongs to the $Y_{\bar{P}}$-saturated family associated to P (cf. definition II.11.3 of [20]). Therefore by the theorem III.3.13 of [20], there is $B_o \in P$ such that $\operatorname{supp} T^\Delta \subset \bar{B}_o^{\upsilon X}$ hence such that $\operatorname{supp} T^\Delta \subset \bar{B}_o^X$ since P is contained in $K(X)$.

Now let us remark that we have

$$\varphi \in C(X;E) \ \& \ \varphi(\operatorname{supp} T^\Delta) = \{0\} \Rightarrow \varphi \in T.$$

This is a direct consequence of the Singer theorem III.3.4 since for $\varphi \in C(X;E)$ such that $\varphi(\operatorname{supp} T^\Delta) = \{0\}$, we have

$$\langle \varphi, \varphi' \rangle = \int \varphi \, dm_{\varphi'} = 0, \quad \forall \varphi' \in T^\Delta,$$

hence $\varphi \in T^{\Delta\nabla} = T$. Therefore we have

$$K(T) \subset \overline{\operatorname{supp} T^\Delta}^X \subset \overline{B_o}^X. \qquad (\star)$$

By the theorem I.8.8, $CRC_u(X;E)$ is topologically isomorphic to $C_c(\beta X;E)$ hence is quasi-barrelled. As T is a bornivorous barrel of $C_P(X;E)$, $T \cap CRC_u(X;E)$ is a bornivorous barrel of $CRC_u(X;E)$ hence a neighborhood of the origin. Therefore there is a continuous semi-norm q on E such that

$$T_q = \{\varphi \in CRC(X;E) : \sup_{x \in X} q(\varphi(x)) \leq 1\} \subset T \cap CRC(X;E).$$

Of course

$$T'_q = \{\varphi \in C(X;E) : \sup_{x \in X} q(\varphi(x)) \leq 1\}$$

is a closed subset of $C_P(X;E)$ and contains T_q. Therefore to prove the equality

$$\overline{T_q}^{C_P(X;E)} = T'_q$$

we just have to establish that T_q is $C_P(X;E)$-dense in T'_q. Let us consider $\psi \in T'_q$, $p \in P$, $\varepsilon > 0$ and $B \in P$. The sets

$$\{y \in X : p(\psi(x) - \psi(y)) < \varepsilon\} \quad (x \in \bar{B}^X)$$

constitute an open covering of the compact subset $\bar{B}^X$ of X from which we can extract a finite covering; let $\{\Omega_1, \ldots, \Omega_J\}$ be such a covering.

By the proposition I.5.2, there are then continuous functions f_j $(j \leq J)$ on X with values in $[0,1]$, such that supp $f_j \subset \Omega_j$ for every $j \leq J$ and $\Sigma_{j=1}^{J} f_j$ is identically 1 on a neighborhood of $\bar{B}^X$ and bounded by 1 on X. For every $j \leq J$, let us choose a point $x_j \in \Omega_j$ and set $\psi(x_j) = e_j$. Then of course $\Sigma_{j=1}^{J} f_j e_j$ belongs to $C(X;E)$ but as

$$\{ \sum_{j=1}^{J} f_j(x) e_j : x \in X\}$$

is a bounded subset of finite dimension of E, it also belongs to $CRC(X;E)$; as a direct consequence, it then belongs to T_q. To conclude this part of the proof, it is enough then to check that we have

$$p_B(\psi - \sum_{j=1}^{J} f_j e_j) \leq \sup_{x \in B} \sum_{j=1}^{J} f_j(x)\, p(\psi(x) - \psi(x_j)) \leq \varepsilon.$$

Therefore we have

$$T \supset \overline{T_q}^{C_P(X;E)} = \{\varphi \in C(X;E) : \sup_{x \in X} q(\varphi(x)) \leq 1\}$$

hence

$$T \supset \{\varphi \in C(X;E) : \sup_{x \in K(T)} q(\varphi(x)) < 1\}$$

by the theorem II.1.4.

The conclusion follows then at once from (*).///

THEOREM IV.6.2. *The space* $C_{Q,s}(X;E)$ *is quasi-barrelled if and only if* (E,Q) *is quasi-barrelled. Moreover the quasi-barrelled space associated to* $C_{Q,s}(X;E)$ *is the space* $C_{Q_e,s}(X;E)$ *where* Q_e *is the system of semi-norms of the quasi-barrelled space associated to* (E,Q).

In particular the space $C_s(X;E)$ *is quasi-barrelled if and only if* E *is quasi-barrelled. Moreover the quasi-barrelled space associated to* $C_s(X;E)$ *is the space* $C_{P_e,s}(X;E)$ *where* P_e *is the system of semi-norms of the quasi-barrelled space associated to* E.

Proof. The condition is necessary. Let B be a bounded subset of

$(E,Q)'_b$ and let us fix an element x of X. Then

$$B_x = \{<.(x),e'> : e' \in B\}$$

is of course a bounded subset of $C_{Q,s}(X;E)'_b$ hence is equicontinuous since $C_{Q,s}(X;E)$ is quasi-barrelled. The conclusion is then a direct consequence of the proposition II.7.6.

The condition is sufficient. Let B be a bounded subset of $C_{Q,s}(X;E)'_b$. By the proposition II.7.5, we know that supp B is finite and that for every $x \in$ supp B,

$$\{e'_{\varphi',x} : \varphi' \in B\}$$

is a bounded subset of $(E,Q)'_b$ hence equicontinuous since (E,Q) is quasi-barrelled. Hence the conclusion by the proposition II.7.6.

Now let us consider the characterization of the quasi-barrelled space associated to $C_{Q,s}(X;E)$. By use of the transfinite construction of the paragraph I.5 of [20], it is sufficient to prove that for every ordinal number α, we have the equality in between the Hausdorff locally convex topological vector spaces

$$C_{Q,s}(X;E)_\alpha \quad \text{and} \quad C_{Q_\alpha,s}(X;E)$$

where Q_α is the system of semi-norms of $(E,Q)_\alpha$. For $\alpha = 1$, this is a direct consequence of the proposition II.7.5 and of the definition of Q_1. In exactly the same way we get that the equality for α implies the equality for $\alpha + 1$. To conclude, it is then enough to remark that if α is a limit ordinal number and if the equality holds for every ordinal number $\beta < \alpha$, then the equality holds also for α by use of the definitions of $C_{Q,s}(X;E)_\alpha$ and Q_α in that case.

The particular case is then immediate.///

<u>Remark</u> IV.6.3. As usual

a) $c_o(E)$ is the Hausdorff locally convex topological vector space obtained by endowing the linear space of the null sequences $\underline{e}$ of E with the uniform convergence, i.e. with the system of semi-norms $P^{(\infty)} = \{p^{(\infty)} : p \in P\}$ where for every $p \in P$, $p^{(\infty)}$ is defined by

$$p^{(\infty)}(\underline{e}) = \sup_m p(e_m), \quad \forall\, \underline{e} \in c_o(E).$$

b) $l_1(E)$ is the Hausdorff locally convex topological vector space obtained by endowing the linear space of the absolutely summable sequences $\underline{e}$ of E with the π-topology, i.e. with the system of semi-norms $P^{(1)} = \{p^{(1)} : p \in P\}$ where for every $p \in P$, $p^{(1)}$ is defined by

$$p^{(1)}(\underline{e}) = \sum_{m=1}^{\infty} p(e_m) < \infty, \quad \forall\ \underline{e} \in l_1(E).$$

The structure of the dual of $c_o(E)$ is well known : a linear functional $\underline{e}'$ on $c_o(E)$ is such that

$$|<\underline{e},\underline{e}'>| \leqslant C\ p^{(\infty)}(\underline{e}), \quad \forall\ \underline{e} \in c_o(E),$$

if and only if there are sequences c_m of $\mathbb{C}$ and d'_m of E' such that

$$\sum_{m=1}^{\infty} |c_m| \leqslant 1,$$

$$|<e,d'_m>| \leqslant C\ p(e), \quad \forall\ m \in \mathbb{N},\ \forall\ e \in E,$$

and

$$<\underline{e},\underline{e}'> = \sum_{m=1}^{\infty} c_m <e_m,d'_m>, \quad \forall\ \underline{e} \in c_o(E).$$

Therefore a sequence $\underline{e}'$ of elements e'_m of E' determines a continuous linear functional on $c_o(E)$ if and only if there is $p \in P$ such that $\Sigma_{m=1}^{\infty} \|e'_m\|_p < \infty$ or, which amounts to the same, if and only if there is an absolutely convex equicontinuous subset B of E' such that $\Sigma_{m=1}^{\infty} p_B(e'_m) < \infty$.

Let us also recall that, following A. Pietsch, E <u>has property</u> (B) if for every bounded subset $\underline{B}$ of $l_1(E)$, there is an absolutely convex bounded subset B of E such that

$$\sum_{m=1}^{\infty} p_B(e_m) \leqslant 1, \quad \forall\ \underline{e} \in \underline{B}.$$

The introduction of the spaces having property (B) is motivated by the following property : <u>if</u> E <u>has property</u> (B), <u>then every absolutely summable sequence of</u> E <u>is totally summable</u>.

Now let us recall the following fundamental examples of E'_b spaces having property (B) :

a) <u>every metrizable space has property</u> (B). In particular, <u>if</u> E <u>has a countable fundamental family of bounded subsets, then</u> E'_b <u>has property</u> (B),

b) <u>if</u> E <u>is metrizable, then</u> E'_b <u>has property</u> (B),

c) <u>if</u> E <u>is quasi-barrelled and nuclear, then</u> E'_b <u>is nuclear hence has property</u> (B).

THE MARQUINA-SANZ SERNA THEOREM IV.6.4.[12]

a) <u>The space</u> $c_o(E)'_b$ <u>is a sequentially dense topological subspace of</u> $l_1(E'_b)$.

b) <u>The space</u> $c_o(E)$ <u>is quasi-barrelled if and only if</u> E <u>is quasi-barrelled and such that</u> E'_b <u>has property</u> (B).

c) <u>If</u> $c_o(E)$ <u>is quasi-barrelled, the spaces</u> $c_o(E)'_b$ <u>and</u> $l_1(E'_b)$ <u>coincide</u>.

<u>Proof</u>. a) On one hand $c_o(E)'_b$ is a sequentially dense subspace of $l_1(E'_b)$ since it contains the finite sequences and since with self-evident notations, for every $\underline{e}' \in c_o(E)'$ and every bounded subset B of E we have

$$\sum_{m=1}^{\infty} \sup_{e\in B} |\langle e, c_m d'_m\rangle| \leq C \sup_{e\in B} p(e). \sum_{m=1}^{\infty} |c_m| < \infty.$$

On the other hand let at first $\underline{B}$ be a bounded subset of $c_o(E)$. Then of course

$$B = \{e_m : m \in \mathbb{N},\ \underline{e} \in \underline{B}\}$$

is a bounded subset of E such that

$$\sup_{\underline{e}\in\underline{B}} |\langle \underline{e}, \underline{e}'\rangle| \leq \sum_{m=1}^{\infty} \sup_{e\in B} |\langle e, c_m d'_m\rangle|, \quad \forall\ \underline{e}' \in c_o(E)',$$

which proves that $c_o(E)'_b$ has a coarser topology than the one induced by $l_1(E'_b)$. Let now B be a bounded subset of E. Then of course

$$\underline{B} = \{\underline{e} \in c_o(E) : e_m \in B, \quad \forall\ m \in \mathbb{N}\}$$

is a bounded subset of $c_o(E)$ such that

$$\sum_{m=1}^{\infty} \sup_{e \in B} |\langle e, c_m d'_m \rangle| \leq \sup_{\underline{e} \in \underline{B}} |\langle \underline{e}, \underline{e}' \rangle|, \quad \forall\, \underline{e}' \in c_o(E)',$$

as one can check easily. This proves that $c_o(E)'_b$ has also a finer topology than the one induced by $l_1(E'_b)$. Hence the conclusion.

b) and c). Let us suppose first that $c_o(E)$ is quasi-barrelled. It is well known that E is topologically isomorphic to a complemented subspace of $c_o(E)$ therefore E is quasi-barrelled. Moreover part a) of this theorem and the fact that $c_o(E)$ is quasi-barrelled imply directly that the spaces $c_o(E)'_b$ and $l_1(E'_b)$ coincide. Finally every bounded subset $\underline{B}$ of $l_1(E'_b) = c_o(E)'_b$ is then equicontinuous on $c_o(E)$: there are $p \in P$ and $C > 0$ such that

$$\sup_{\underline{e}' \in \underline{B}} |\langle ., \underline{e}' \rangle| \leq C\, p^{(\infty)}(.) \quad \text{on} \quad c_o(E).$$

Therefore with self evident notations, the absolutely convex hull B of

$$\{d'_m(\underline{e}') : \underline{e}' \in \underline{B},\ m \in \mathbb{N}\}$$

is equicontinuous on E and such that

$$\sum_{m=1}^{\infty} p_B(e'_m) \leq \sum_{m=1}^{\infty} |c_m(\underline{e}')| \leq 1, \quad \forall\, \underline{e}' \in \underline{B},$$

hence E'_b has property (B).

Finally let us suppose that E is quasi-barrelled and that E'_b has property (B). As E'_b has property (B), every absolutely summable sequence of E'_b is totally summable; this means that for every $\underline{e}' \in l_1(E'_b)$, there is a bounded subset B of E'_b such that

$$\sum_{m=1}^{\infty} p_B(e'_m) < +\infty .$$

As E is quasi-barrelled, B is equicontinuous on E and therefore $\underline{e}'$ belongs to $c_o(E)'$. By use of part a), this proves that the spaces $c_o(E)'_b$ and $l_1(E'_b)$ coincide. Now let $\underline{B}$ be a bounded subset of $c_o(E)'_b$. It is then bounded in $l_1(E'_b)$ and as E'_b has property (B), there is a bounded subset B of E'_b such that

$$\sum_{m=1}^{\infty} p_B(e'_m) \leq 1, \quad \forall\ \underline{e}' \in \underline{B}.$$

As E is quasi-barrelled, B is equicontinuous. It is then immediate to check that $\underline{B}$ is equicontinuous on $c_o(E)$.///

Remark IV.6.5. Here is the proof of Marquina and Sanz Serna of the example IV.2.4. It is enough to prove that for every set J, $(\oplus_{j\in J} \mathbb{C})'_b = \mathbb{C}^J$ has property (B) if and only if J is countable. The condition is necessary since the family $(\varepsilon_j)_{j\in J}$ of $\mathbb{C}^J$ is certainly absolutely summable hence totally summable which implies that it contains at most countably many non zero elements. The condition is of course sufficient since then $\mathbb{C}^J$ is metrizable.

PROPOSITION IV.6.6 [15] The following are equivalent

(a) there is an infinite Hausdorff compact space K_o such that $C_c(K_o;E)$ is quasi-barrelled,

(b) for every Hausdorff compact space K, $C_c(K;E)$ is quasi-barrelled,

(c) E is quasi-barrelled and E'_b has property (B).

Proof. (a) ⇒ (b). Let K be a Hausdorff compact space. Moreover let T be a bornivorous barrel of $C_c(K;E)$. Then we denote by D the family of all the absolutely convex subsets D of E such that $C(K;D) \subset T$ and we set

$$T_o = \bigcup_{D\in \mathcal{D}} C(K_o;D).$$

Now we prove that this subset T_o of $C_c(K_o;E)$ is

i) absolutely convex. In fact, for every $D_1, D_2 \in D$ and every $r \in [0,1]$, the part c) of the proposition I.6.3 implies

$$C(K;rD_1 + (1-r)D_2) \subset \overline{rC(K;D_1) + (1-r)C(K;D_2)}$$

where the second member is contained in T. The conclusion is then direct since this implies $rD_1 + (1-r)D_2 \in D$.

ii) bornivorous. For every bounded subset B_o of $C_c(K_o;E)$, the absolutely convex hull B of

$$\{\varphi(x) : \varphi \in B_o, x \in K_o\}$$

is a bounded subset of E. Therefore C(K;B) is an absolutely convex bounded subset of $C_c(K;E)$ hence is absorbed by T : there is $r > 0$ such that

$$C(K;rB) = rC(K;B) \subset T.$$

The conclusion is then direct since we certainly have

$$B_o \subset C(K_o;B).$$

iii) closed. Let φ_o be an element of $\overline{T_o}$ and denote by D_{φ_o} the absolutely convex hull of $\varphi_o(K_o)$. To conclude, it is then enough to prove that D_{φ_o} belongs to $\mathcal{D}$. As T is a barrel, by the part b) of proposition I.6.2, it is enough to prove that we have $P(K;D_{\varphi_o}) \subset T$. To do this, let $\{g_1, \ldots, g_J\}$ be a finite continuous partition of unity on K and let $e_1, \ldots, e_J$ be elements of D_{φ_o}. There are then

$$L \in \mathbb{N}, \{c_{j,l} : j \leq J, l \leq L\} \subset \mathbb{C} \text{ and } \{x_{j,l} : j \leq J, l \leq L\} \subset K_o$$

such that

$$\sum_{l=1}^{L} |c_{j,l}| \leq 1, \quad \forall j \leq J,$$

and

$$\sum_{l=1}^{L} c_{j,l}\, \varphi_o(x_{j,l}) = e_j, \quad \forall j \leq J .$$

As φ_o belongs to $\overline{T_o}$, for every $p \in P$ and $r > 0$, there are $D \in \mathcal{D}$ and $\psi_o \in C(K_o;D)$ such that

$$p_{K_o}(\varphi_o - \psi_o) \leq r$$

hence such that

$$p(\varphi_o(x_{j,l}) - \psi_o(x_{j,l})) \leq r, \quad \forall j \leq J, \forall l \leq L.$$

Therefore

$$\psi = \sum_{j=1}^{J} g_j \sum_{l=1}^{L} c_{j,l}\, \psi_o(x_{j,l})$$

belongs to $C(K;D)$ hence to T and is such that

$$p_K\Big(\sum_{j=1}^{J} g_j\, e_j - \psi\Big) \leqslant r$$

since for every $y \in K$, we have

$$p\Big(\sum_{j=1}^{J} g_j(y) e_j - \psi(y)\Big)$$

$$\leqslant \sum_{j=1}^{J} g_j(y) \sum_{l=1}^{L} |c_{j,l}|\; p(\varphi_o(x_{j,l}) - \psi_o(x_{j,l})) \leqslant r.$$

This proves that $\Sigma_{j=1}^{J} g_j e_j$ belongs to the closure of T hence to T.

Being a bornivorous barrel of $C_c(K_o;E)$, T_o is a neighborhood of the origin : there are $p \in P$ and $r > 0$ such that

$$C(K_o;b_p(r)) = \{\varphi \in C(K_o;E) : p_{K_o}(\varphi) \leqslant r\} \subset T_o.$$

To conclude, it is then enough to prove that this implies the inclusion $C(K;b_p(r)) \subset T$.

As T is a barrel, it is enough to prove that we have $P(K;b_p(r)) \subset T$. To do this, let $\{g_1, \ldots, g_J\}$ be a finite continuous partition of unity on K and let $e_1, \ldots, e_J$ be elements of $b_p(r)$. As K_o is infinite, we can choose distinct points $x_1, \ldots, x_J$ of K_o and a finite continuous partition of unity $\{f_1, \ldots, f_J\}$ on K_o such that $f_j(x_k) = \delta_{j,k}$ for every $j, k \leqslant J$. Then we have

$$\sum_{j=1}^{J} f_j\, e_j \in C(K_o;b_p(r)) \subset T_o$$

and this implies the existence of $D \in \mathcal{D}$ such that

$$\sum_{j=1}^{J} f_j\, e_j \in C(K_o;D)$$

hence such that $\{e_1, \ldots, e_J\} \subset D$ which implies

$$\sum_{j=1}^{J} g_j e_j \in C(K;D).$$

Hence the conclusion.

(b) ⇒ (c) and (c) ⇒ (a) are direct consequences of the Marquina-Sanz Serna theorem IV.6.4 and of the following result which has its own interest.///

THEOREM IV.6.7. *Let* $\alpha\mathbb{N}$ *be the Alexandroff compactification of the discrete space* $\mathbb{N}$. *Then the linear map*

$$T : C_c(\alpha\mathbb{N};E) \rightarrow c_o(E)$$

defined by

$$(T\varphi)_m = \begin{cases} \varphi(\infty) & \text{if } m = 1 \\ \varphi(m-1) - \varphi(\infty) & \text{if } m > 1 \end{cases}$$

is a topological isomorphism.

Proof. This is immediate and well known.///

THE MENDOZA THEOREM IV.6.8 [15]

a) *If* X *is pseudo-finite* (i.e. every compact subset of X is finite), *then* $C_c(X;E)$ *is quasi-barrelled if and only if* E *is quasi-barrelled.*

b) *If* X *is not pseudo-finite, then* $C_c(X;E)$ *is quasi-barrelled if and only if* $C_c(X)$ *and* E *are quasi-barrelled and such that* E_b' *has property* (B).

Proof. a) is a particular case of the theorem IV.6.2.

b) The condition is necessary. By the theorem IV.2.2, we know already that $C_c(X)$ and E are quasi-barrelled spaces. Now let K be an infinite compact subset of X. To conclude by the proposition IV.6.5, it is enough to prove that $C_c(K;E)$ is quasi-barrelled. The restriction map

$$R_K : C_c(X;E) \rightarrow C_c(K;E)$$

is of course a linear homomorphism. Therefore $R_K C_c(X;E)$ is a quasi-barrelled linear subspace of $C_c(K;E)$. As it contains $C(K)\otimes E$, $R_K C_c(X;E)$ is also dense in $C_c(K;E)$. Hence the conclusion.

The sufficiency of the condition is a direct consequence of the theorems IV.6.1 and IV.6.6.///

At this stage it is quite natural to look for examples of quasi-barrelled spaces E such that E'_b has property (B) and for stability conditions relative to these spaces.

EXAMPLES IV.6.9. The examples of spaces E such that E'_b has property (B) that we have recalled in the remark IV.6.3 lead immediately to the following results : *examples of quasi-barrelled spaces* E *such that* E'_b *has property* (B) *are given by*

a) *the metrizable spaces*,

b) *the quasi-barrelled spaces having a countable fundamental family of bounded subsets*,

c) *the quasi-barrelled and nuclear spaces*.

STABILITY PROPERTIES IV.6.10 [15]

a) *If the spaces* E_j $(j\in J)$ *are quasi-barrelled and if their strong duals have property* (B), *then their product is quasi-barrelled and its strong dual has property* (B).

b) *If* E *is quasi-barrelled and if* E'_b *has property* (B), *then every topologically complemented subspace of* E *is quasi-barrelled and its strong dual has property* (B).

c) *If a dense topological subspace of* E *is quasi-barrelled and if its strong dual has property* (B), *then* E *is quasi-barrelled and* E'_b *has property* (B).

d) *If the convention* I.7.1 *holds, if each* E_m *is quasi-barrelled and if each* $(E_m)'_b$ *has property* (B), *then* Ind E_m *is quasi-barrelled and its strong dual has property* (B).

e) *If* $C_c(X)$ *and* E *are quasi-barrelled and if* E'_b *has property* (B), *then* $C_c(X;E)$ *is quasi-barrelled and its strong dual has property* (B).

Proof. a) and b) are direct consequences of the proposition IV.6.6 and of the fact that for every Hausdorff compact space K, the spaces

$$C_c(K; \prod_{j\in J} E_j) \quad \text{and} \quad \prod_{j\in J} C_c(K;E_j)$$

are topologically isomorphic.

c) Let us denote by L the linear subspace. Then for every Hausdorff compact space K, $C_c(K;L)$ is quasi-barrelled and is, by the theorem I.5.3, dense in $C_c(K;E)$; this implies that $C_c(K;E)$ is quasi-barrelled.

d) By the Mujica theorem I.7.2, for every Hausdorff compact space K, the quasi-barrelled space Ind $C_c(K;E_m)$ is a dense topological space of $C_c(K;\text{Ind } E_m)$; this implies that $C_c(K;\text{Ind } E_m)$ is quasi-barrelled.

e) For every Hausdorff compact space K, $C_c(X\times K)$ is a quasi-barrelled space and therefore $C_c(X\times K;E)$ is quasi-barrelled. Hence the conclusion since $C_c(X\times K;E)$ is then topologically isomorphic to a dense quasi-barrelled topological subspace of $C_c(K;C_c(X;E))$.///

IV.7. BARRELLED $C_P(X;E)$ SPACES (SUFFICIENT CONDITION)

THE MENDOZA THEOREM IV.7.1 [13] If P is contained in $K(X)$ and if $C_P(X)$ and $C_c(\beta X;E)$ are barrelled, then $C_P(X;E)$ is barrelled.

Proof. A direct simplification of the proof of the theorem IV.6.1 gives the result.///

THEOREM IV.7.2. The space $C_{Q,s}(X;E)$ is barrelled if and only if $C_s(X)$ and (E,Q) are barrelled. Moreover if $C_s(X)$ is barrelled, the barrelled space associated to $C_{Q,s}(X;E)$ is the space $C_{Q_t,s}(X;E)$ where Q_t is the system of semi-norms of the barrelled space associated to (E,Q).

In particular the space $C_s(X;E)$ is barrelled if and only if $C_s(X)$ and E are barrelled. Moreover if $C_s(X)$ is barrelled, the barrelled space associated to $C_s(X;E)$ is the space $C_{P_t,s}(X;E)$ where P_t is the system of semi-norms of the barrelled space associated to E.

Proof. The first part of the result is a direct consequence of the following two propositions which have their own interest.

PROPOSITION IV.7.3. The set supp T^Δ is finite for every barrel T

<u>of</u> $C_{Q,s}(X;E)$ <u>if and only if</u> $C_s(X)$ <u>is barrelled</u>.

<u>Proof</u>. The condition is necessary. If $C_s(X)$ is not barrelled, we know by the proposition III.3.13 of [20] that there is a bounding subset B of X which is not finite. Then for p ∈ P different from 0,

$$T = \{\varphi \in C(X;E) : p_B(\varphi) \leq 1\}$$

$$= \bigcap_{x \in B} \{\varphi \in C(X;E) : p(\varphi(x)) \leq 1\}$$

is a barrel of $C_s(X;E)$ hence of $C_{Q,s}(X;E)$. However it is clear that supp T^{Δ} must contain B. Hence a contradiction.

The condition is sufficient. In fact if T is a barrel of $C_{Q,s}(X;E)$, its polar T^{Δ} is a bounded subset of $C_{Q,s}(X;E)'_s$. By the proposition II.7.4, supp T^{Δ} is then a bounding subset of X, hence a finite subset of X since $C_s(X)$ is barrelled.///

PROPOSITION IV.7.4. <u>If</u> $C_s(X)$ <u>is barrelled, then the space</u> $C_{Q,s}(X;E)$ <u>is barrelled if and only if</u> (E,Q) <u>is barrelled</u>.

<u>Proof</u>. The condition is necessary. Let B be any bounded subset of $(E,Q)'_s$ and fix a point $x_o \in X$. Then it is trivial that

$$\{\langle .(x_o), e' \rangle : e' \in B\}$$

is a bounded subset of $C_{Q,s}(X;E)'_s$ and that its support is equal to $\{x_o\}$. Hence the conclusion by the proposition II.7.6.

The condition is sufficient. Let B be any bounded subset of $C_{Q,s}(X;E)'_s$. By proposition II.7.4, supp B is a bounding subset of X hence a finite subset of X since $C_s(X)$ is barrelled. To conclude by the proposition II.7.6, it is then enough to prove that for every $x \in \text{supp } B$,

$$\{e'_{\varphi',x} : \varphi' \in B\}$$

is a bounded subset of $(E,Q)'_s$. To get this, we choose a function $f \in C(X)$ such that $f(y) = \delta_{x,y}$ for every $y \in \text{supp } B$. Then we have

$$\sup_{\varphi' \in B} |\langle e, e'_{\varphi',x} \rangle| = \sup_{\varphi' \in B} |\langle fe, \varphi' \rangle| < \infty, \quad \forall\, e \in E,$$

hence the conclusion.///

Now let us prove the part of the theorem IV.7.2 related to the characterization of the barrelled space associated to $C_{Q,s}(X;E)$.

By use of the transfinite construction of the paragraph I.4 of [20], it is sufficient to prove that for every ordinal number α, we have the equality in between the Hausdorff locally convex topological vector spaces

$$C_{Q,s}(X;E)_\alpha \quad \text{and} \quad C_{Q_\alpha,s}(X;E)$$

where Q_α is the system of semi-norms of $(E,Q)_\alpha$. For $\alpha = 1$, this is a direct consequence of the proposition IV.7.3 since a subset B of $C_{Q,s}(X;E)'_s$ is bounded if and only if the following two conditions are fulfilled

a) supp B is a finite subset of X [because $C_s(X)$ is barrelled],

b) for every $x \in$ supp B,

$$B_x = \{e'_{\varphi',x} \quad : \quad \varphi' \in B\}$$

is a bounded subset of $(E,Q)'_s$ [because we have a)].
In exactly the same way we get that the equality for α implies the equality for $\alpha + 1$. To conclude, it is then enough to remark that if α is a limit ordinal number and if the equality holds for every ordinal number $\beta < \alpha$, then the equality holds also for α by use of the definitions of $C_{Q,s}(X;E)_\alpha$ and Q_α in that case.

The particular case is then immediate.

Hence the conclusion of the theorem IV.7.2.///

In the characterization of the barrelled $C_c(X;E)$ spaces, the spaces $c_o(E)$ are very important. The following result will lead us to a characterization of the barrelled $c_o(E)$ spaces; it is generalized at the proposition IV.7.8.

PROPOSITION IV.7.5 [16] Every bounded subset of E'_s is bounded in E'_b if and only if every bounded subset of $c_o(E)'_s$ is bounded in $c_o(E)'_b$.

Proof. The condition is of course sufficient since E is topologically isomorphic to a topologically complemented subspace of $c_o(E)$; let us establish by contradiction that it is necessary.

Let B be a bounded subset of $c_o(E)'_s$ and let us suppose that B is not a bounded subset of $c_o(E)'_b$.

On one hand by the part a) of the Marquina-Sanz Serna theorem IV.6.4, there is then an absolutely convex and bounded subset B of E such that

$$\sup_{\underline{e}' \in B} \sum_{m=1}^{\infty} \sup_{e \in B} |\langle e, e'_m \rangle| = +\infty. \qquad (\star)$$

On the other hand for every $m \in \mathbb{N}$,

$$\{e'_m : \underline{e}' \in B\}$$

is certainly a bounded subset of E'_s hence a bounded subset of E'_b : we have

$$\sup_{\underline{e}' \in B} \sup_{e \in B} |\langle e, e'_m \rangle| = C_m < +\infty, \quad \forall\, m \in \mathbb{N}. \qquad (\star\star)$$

Let us set $M_m = \Sigma_{k=1}^m C_k$ for every $m \in \mathbb{N}$.

By use of $(\star)$, there is $\underline{e}'^{(1)} \in B$ such that

$$\sum_{m=1}^{\infty} \sup_{e \in B} |\langle e, e'^{(1)}_m \rangle| > 1$$

and therefore $n_1 \in \mathbb{N}$ such that

$$\sum_{m=1}^{n_1} \sup_{e \in B} |\langle e, e'^{(1)}_m \rangle| > 1 \quad \text{and} \quad \sum_{m=n_1+1}^{\infty} \sup_{e \in B} |\langle e, e'^{(1)}_m \rangle| < 1.$$

By use of $(\star)$ and $(\star\star)$, we certainly have then

$$\sup_{\underline{e}' \in B} \sum_{m=n_1+1}^{\infty} \sup_{e \in B} |\langle e, e'_m \rangle| = +\infty.$$

In this way we can construct a strictly increasing sequence $n_j \in \mathbb{N}$ and a sequence $\underline{e}'^{(j)} \in B$ such that

$$\sum_{m=n_{j-1}+1}^{n_j} \sup_{e \in B} |\langle e, e'^{(j)}_m \rangle| > j(j + M_{n_{j-1}} + 1)$$

and

$$\sum_{m=n_j+1}^{\infty} \sup_{e\in B} |<e, e_m'^{(j)}>| < 1$$

for every $j \geq 2$.

Therefore there is a sequence e_m of B such that

$$\sum_{m=1}^{n_1} <e_m, e_m'^{(1)}> > 1$$

and

$$\sum_{m=n_{j-1}+1}^{n_j} <e_m, e_m'^{(j)}> > j(j + M_{n_{j-1}} + 1)$$

for every integer $j \geq 2$.

Now let us set

$$d_m = \begin{cases} e_m & \text{if } 1 \leq m \leq n_1 \\ \frac{1}{j} e_m & \text{if } n_{j-1} < m \leq n_j \text{ and if } j \geq 2. \end{cases}$$

Of course this sequence d_m belongs to $c_o(E)$. However for every $j \geq 2$, we also have

$$\left| \sum_{m=1}^{\infty} <d_m, e_m'^{(j)}> \right|$$

$$\geq \left| \sum_{m=n_{j-1}+1}^{n_j} <d_m, e_m'^{(j)}> \right| - \sum_{m=1}^{n_{j-1}} |<d_m, e_m'^{(j)}>| - \sum_{m=n_j+1}^{\infty} |<d_m, e_m'^{(j)}>|$$

$$\geq \frac{1}{j} \sum_{m=n_{j-1}+1}^{n_j} <e_m, e_m'^{(j)}> - M_{n_{j-1}} - 1$$

$$\geq j.$$

Hence a contradiction.///

THEOREM IV.7.6 [16] <u>The space</u> $c_o(E)$ <u>is barrelled if and only if</u> E <u>is barrelled and such that</u> E_b' <u>has property</u> (B).

<u>Proof</u>. By the part b) of the Marquina-Sanz Serna theorem IV.6.4, we know that $c_o(E)$ is quasi-barrelled if and only if E is quasi-barrelled and such that E_b' has property (B). Therefore the necessity of the condition is a direct consequence of the well known fact E is topologically isomorphic to a complemented subspace of $c_o(E)$ and the sufficiency of the condition is an immediate consequence of the previous result.///

PROPOSITION IV.7.7 [15] <u>The following are equivalent</u>

(a) <u>there is an infinite Hausdorff compact space</u> K_o <u>such that</u> $C_c(K_o;E)$ <u>is barrelled</u>,

(b) <u>for every Hausdorff compact space</u> K, $C_c(K;E)$ <u>is barrelled</u>,

(c) E <u>is barrelled and</u> E_b' <u>has property</u> (B).

<u>Proof</u>. (b) ⇒ (c) is a direct consequence of the theorem IV.2.2 and of the proposition IV.6.6.

(c) ⇒ (a) is a direct consequence of the theorem IV.6.7 and of the preceding theorem.

(a) ⇒ (b) By the theorem IV.2.2, E is barrelled and by the theorem IV.6.6, we know that for every Hausdorff compact space K, $C_c(K;E)$ is quasi-barrelled. The conclusion follows then at once from the following result which generalizes the proposition IV.7.5.///

PROPOSITION IV.7.8 [15] <u>Every bounded subset of</u> E_s' <u>is bounded in</u> E_b' <u>if and only if, for every Hausdorff compact space</u> K, <u>every bounded subset of</u> $C_c(K;E)_s'$ <u>is bounded in</u> $C_c(K;E)_b'$.

<u>Proof</u>. The sufficiency of the condition is a direct consequence of the theorem I.4.4. Let us establish its necessity in two steps.

a) Let us first prove that it is enough to establish that the result holds for metrizable compact spaces.

Let K be a Hausdorff compact space and let $\mathcal{B}$ be a bounded subset of $C_c(K;E)_s'$ which is not bounded in $C_c(K;E)_b'$. It is clear that the sets C(K;B) where B is any absolutely convex bounded subset of E constitute a fundamental family of the bounded subsets of $C_c(K;E)$. Therefore there is an absolutely convex bounded subset B_o of E such that $\mathcal{B}$

is not uniformly bounded on $C(K;B_o)$ nor on $P(K;B_o)$, by use of the part b) of the proposition I.6.2. There is then a sequence $\varphi_m \in P(K;B_o)$ on which B is not uniformly bounded. Each φ_m takes its values in a finite dimensional subspace of E hence in a metrizable subspace of E. Therefore the coarsest topology on K such that all the φ_m are continuous is pseudo-metrizable for a pseudo-metric d such that, with standard notations,

$$\tilde{K} = (K/d, \tilde{d})$$

is a metrizable compact space. Under these circumstances, the canonical map

$$I : C_c(\tilde{K};E) \rightarrow C_c(K;E)$$

is of course linear and continuous. Therefore

$$\tilde{B} = \{\varphi' \circ I : \varphi' \in B\}$$

is a bounded subset of $C_c(\tilde{K};E)'_s$ and not a bounded subset of $C_c(\tilde{K};E)'_b$ since the sequence φ_m is of course the image by I of a bounded sequence of $C_c(\tilde{K};E)$.

b) To conclude, it is then enough to prove by contradiction that the condition is necessary if moreover the Hausdorff compact space K is metrizable.

Let B be a bounded subset of $C_c(K;E)'_s$ which is not bounded in $C_c(K;E)'_b$. As in part a) of this proof, there is then an absolutely convex bounded subset B_o of E such that B is not uniformly bounded on $P(K;B_o)$.

Let d be a metric on K compatible with the topology of K and such that diam $K \leqslant 1$.

We prove now by recursion the existence of a decreasing sequence Ω_m of open subsets of K such that

$$\operatorname{diam} \Omega_m \leqslant \frac{1}{m}$$

and that B is not uniformly bounded on any of the sets

$$A'_m = \{ \sum_{(k)} f_k e_k : f_k \in C(K), 0 \leq f_k, \text{ supp } f_k \subset \Omega_m,$$

$$e_k \in B_o \text{ for every } k \ \& \ \sum_{(k)} f_k \leq \chi_{\Omega_m} \}.$$

As we have diam $K \leq 1$ and as B is not uniformly bounded on $P(K;B_o)$, we can set $\Omega_1 = K$. Now if $\Omega_1, \ldots, \Omega_{m-1}$ are determined, as $\overline{\Omega_{m-1}}^K$ is a compact subspace of K, we can obtain easily a finite open covering $\{\Omega_{m,j} : j \leq J\}$ of Ω_{m-1} by subsets of Ω_{m-1} of diameter less than or equal to $1/m$. We note then that in the representation $\Sigma_{(k)} f_k e_k$ of each $\varphi \in A'_{m-1}$, we may suppose that each one of the f_k has its support contained in one of the sets $\Omega_{m,j}$ ($j \leq J$) (this is possible since each f_k has a compact support contained in Ω_{m-1} - use proposition I.5.2.). Then it is direct matter to check that one of the $\Omega_{m,j}$ ($j \leq J$) can be chosen as Ω_m.

Of course, the intersection of the Ω_m is either void or reduced to one point. If it is void, we set $A_m = A'_m$ for every m. If it is equal to $\{x_o\}$, we set

$$A_m = \{ \sum_{(k)} f_k e_k : f_k \in C(K), f_k(x_o) = 0, \text{ supp } f_k \subset \Omega_m,$$

$$e_k \in B_o \text{ for every } k \ \& \ \sum_{(k)} |f_k| \leq 2\chi_{\Omega_m} \}$$

for every m. In both cases we get that for every m, B is not uniformly bounded on A_m. This is clear in the first case. In the second let us fix $g_m \in C(K)$ such that $0 \leq g_m \leq \chi_K$, $g_m(x_o) = 1$ and supp $g_m \subset \Omega_m$. Then for every $\varphi \in A'_m$, we have

$$\varphi = g_m \varphi(x_o) + [\varphi - g_m \varphi(x_o)]$$

with $\varphi - g_m \varphi(x_o) \in A_m$ as one can check directly. To conclude it is then sufficient to note that

$$\{\langle g_m \cdot, \varphi' \rangle : \varphi' \in B\}$$

is certainly a bounded subset of E'_s hence of E'_b, which implies that B is uniformy bounded on

$$\{g_m \varphi(x_o) : \varphi \in A'_m\}.$$

Therefore, setting $M_o = 0$, there is a strictly increasing sequence $M_m \in \mathbb{N}$ and sequences $\varphi'_m \in B$, $e_m^{(o)} \in B_o$ and $f_m \in C(K)$ such that

$$0 \leq f_m \quad \text{or} \quad f_m(x_o) = 0 \text{ depending on the case,}$$

$$\operatorname{supp} f_k \subset \Omega_m \quad \text{if} \quad M_{m-1} < m \leq M_m,$$

$$\sum_{k=M_{m-1}+1}^{M_m} |f_k| \leq 2\chi_{\Omega_m}$$

$$< \sum_{k=M_{m-1}+1}^{M_m} f_k e_k^{(o)}, \varphi'_m > \; > 2^{2m}$$

for every $m \in \mathbb{N}$.

Now let us consider the map

$$S : c_o(E) \rightarrow C_c(K;E)$$

defined by

$$S\,\underline{e} = \sum_{m=1}^{\infty} 2^{-m} \sum_{k=M_{m-1}+1}^{M_m} f_k e_k;$$

it is direct matter to check that S is defined, linear and continuous. Therefore the image $S^{\star}B$ of B by the adjoint of S is a bounded subset of $c_o(E)'_s$ hence of $c_o(E)'_b$, by use of the proposition IV.7.5. Hence a contradiction since we have

$$< \sum_{k=M_{m-1}+1}^{M_m} e_k^{(o)} \varepsilon_k, S^{\star}\varphi'_m > \; = \; <2^{-m} \sum_{k=M_{m-1}+1}^{M_m} f_k e_k^{(o)}, \varphi'_m > \; > 2^m$$

for every $m \in \mathbb{N}$.///

THEOREM IV.7.9.[15]

a) <u>If</u> X <u>is pseudo-finite, then</u> $C_c(X;E)$ <u>is barrelled if and only if</u> $C_c(X)$ <u>and</u> E <u>are barrelled</u>.

b) <u>If</u> X <u>is not pseudo-finite, then</u> $C_c(X;E)$ <u>is barrelled if and only if</u> $C_c(X)$ <u>and E are barrelled and such that</u> E'_b <u>has property</u> (B).

Proof. a) is a particular case of the theorem IV.7.2.

b) The necessity of the condition is a direct consequence of the theorems IV.2.2 and IV.6.8. Its sufficiency is a direct consequence of the theorem IV.7.1 and of the proposition IV.7.7.///

EXAMPLES IV.7.10. As in the examples IV.6.9, we get the following results : *examples of barrelled spaces* E *such that* E'_b *has property* (B) *are given by*

a) *the barrelled and metrizable spaces*,

b) *the barrelled spaces having a countable fundamental family of bounded subsets*,

c) *the barrelled and nuclear spaces*.

STABILITY PROPERTIES IV.7.11 [15]

a) *If the spaces* E_j $(j \in J)$ *are barrelled and if their strong duals have property* (B), *then their product is barrelled and its strong dual has property* (B).

b) *If* E *is barrelled and if* E'_b *has property* (B), *then every topologically complemented subspace of* E *is barrelled and its strong dual has property* (B).

c) *If a dense topological subspace of* E *is barrelled and if its strong dual has property* (B), *then* E *is barrelled and its strong dual has property* (B).

d) *If the convention* I.7.1 *holds, if each* E_m *is barrelled and if each* $(E_m)'_b$ *has property* (B), *then* $\text{Ind}\, E_m$ *is barrelled and its strong dual has property* (B).

e) *If* $C_c(X)$ *and* E *are barrelled and if* E'_b *has property* (B), *then* $C_c(X;E)$ *is barrelled and its strong dual has property* (B).

Proof. It is direct by use of the stability properties IV.6.10 and of the stability properties of the barrelled spaces.///

IV.8. EXISTENCE OF A COUNTABLE FUNDAMENTAL FAMILY OF BOUNDED SUBSETS IN $C_p(X;E)$

THEOREM IV.8.1 [20;58] *The following are equivalent*

a) $C_P(X)$ <u>has a countable fundamental family of bounded subsets</u>,

b) <u>if</u> Ω_m <u>is a sequence of open, non void and mutually disjoint subsets of</u> υX, <u>there is</u> $B \in P$ <u>such that</u> $\{m \in \mathbb{N} : \Omega_m \cap B \neq \emptyset\}$ <u>is not finite</u>,

c) <u>every Cauchy sequence of</u> $C_P(X)$ <u>is uniformly Cauchy on</u> X,

d) X <u>is pseudo-compact and</u> $C_P(X)$ <u>is sequentially complete</u>,

e) $B_m = \{f \in C(X) : |f|_X \leq m\}$ <u>is a fundamental family of the bounded subsets of</u> $C_P(X)$.

<u>Proof</u>. (a) ⇒ (b). If it is not the case, there is a sequence Ω_m of open, non void and mutually disjoint subsets of υX such that $\{m \in \mathbb{N} : \Omega_m \cap B\}$ is finite for every $B \in P$. For $m \in \mathbb{N}$, let us fix $x_m \in \Omega_m \cap (\cup_{B \in P} B)$ and choose $f_m \in C(X)$ such that $f_m(x_m) = 1$, $f_m(X) \subset [0,1]$ and $\operatorname{supp} f_m \subset \Omega_m$. Now let $\{B_m : m \in \mathbb{N}\}$ be a fundamental family of the bounded subsets of $C_P(X)$ and set

$$r_m = \sup\{|f(x_m)| : f \in B_m\}, \quad \forall\, m \in \mathbb{N}.$$

It is then easy to check that the sequence $(1 + r_m)f_m$ is bounded in $C_P(X)$ but is not contained in any of the sets B_m. Hence a contradiction.

(b) ⇒ (c). Let f_m be a sequence of $C_P(X)$ which is not uniformly Cauchy on X. There are then $\varepsilon > 0$, strictly increasing sequence r_m and s_m of $\mathbb{N}$ and a sequence x_m of $\cup\,\{B : B \in P\}$ such that

$$|f_{r_m}(x_m) - f_{s_m}(x_m)| > \varepsilon, \quad \forall\, m \in \mathbb{N}.$$

By use of the lemma II.11.6 of [20], there is then a subsequence x_{k_m} of x_m and mutually disjoint open neighborhoods Ω_m of the x_m in υX such that

$$|f_{r_{k_m}}(x) - f_{s_{k_m}}(x)| > \varepsilon, \quad \forall\, m \in \mathbb{N},\ \forall\, x \in \Omega_m.$$

By (b) there is then $B \in P$ such that $\{m \in \mathbb{N} : \Omega_m \cap B \neq \emptyset\}$ is not finite which implies that the sequence f_m is not uniformly Cauchy on B hence is not Cauchy in $C_P(X)$. Hence the conclusion.

(c) ⇒ (d). Of course (c) implies that $C_P(X)$ is sequentially complete. Moreover, for every $f \in C(X)$, it is trivial that the sequence $\theta_m \circ f$ converges in $C_P(X)$ to f; therefore if (c) holds, this convergence is uniform on X, which implies that f is bounded on X. [Let us recall that (cf. III.1.1. of [20]), for $r > 0$, θ_r is the element of $C(\mathbb{C})$ defined by $\theta_r(z) = z$ if $|z| \leq r$ and $\theta_r(z) = rz/|z|$ if $|z| > r$.]

(d) ⇒ (e). Of course these sets B_m are bounded subsets of $C_P(X)$. Let us suppose now that the bounded subset B of $C_P(X)$ is contained in none of the B_m ($m \in \mathbb{N}$). There are then sequences $f_m \in B$ and $x_m \in \cup \{B : B \in P\}$ such that $|f_m(x_m)| \geq m^3$. Of course the series $\Sigma_{m=1}^{\infty} m^{-2}|f_m|$ is Cauchy in $C_P(X)$ hence converges in $C_P(X)$. Hence a contradiction because its limit is not bounded on X since we have

$$\sum_{m=1}^{M} m^{-2}|f_m(x_l)| \geq 1, \quad \forall\, l \in \mathbb{N},\ \forall\, M \geq l.$$

(e) ⇒ (a) is trivial.///

THEOREM IV.8.2.[14] <u>The space</u> $C_P(X;E)$ <u>has a countable fundamental family of bounded subsets if and only if each of the spaces</u> $C_P(X)$ <u>and</u> E <u>has such a family</u>.

<u>Proof</u>. The condition is necessary. In fact, if $\{B_m : m \in \mathbb{N}\}$ is a countable fundamental family of bounded subsets in $C_P(X;E)$, then the sets

$$B_{E,m} = \{e \in E : \chi_X e \in B_m\} \quad (m \in \mathbb{N})$$

and, for any non zero element e_o of E,

$$B_{C,m} = \{f \in C_P(X) : fe_o \in B_m\} \quad (m \in \mathbb{N})$$

constitute fundamental families of bounded subsets in E and $C_P(X)$ respectively. This is direct since the sets $B_{E,m}$ and $B_{C,m}$ ($m \in \mathbb{N}$) are certainly bounded in E and $C_P(X)$ respectively and since for instance, for every bounded subset B_E of E, $B = \{\chi_X e : e \in B_E\}$ is clearly a bounded subset of $C_P(X;E)$: this implies the existence of $m \in \mathbb{N}$ such that $B \subset B_m$ hence such that $B_E \subset B_{E,m}$.

The condition is sufficient. Let $\{B_m : m \in \mathbb{N}\}$ be a fundamental family of bounded subsets in E. Of course we may suppose that the sets B_m are absolutely convex. We prove then that the sets $C(X;B_m)$ ($m \in \mathbb{N}$) constitute a fundamental family of bounded subsets in $C_P(X;E)$. If it is not the case, there is a bounded subset B of $C_P(X;E)$ which is contained in none of the $C(X;B_m)$. There are then sequences φ_m of B and x_m of X such that we have $\varphi_m(x_m) \notin B_m$ for every $m \in \mathbb{N}$. Therefore the sequence $\varphi_m(x_m)$ is not bounded in E which implies the existence of $p \in P$ such that the sequence $p(\varphi_m)$ is not uniformly bounded on X. Hence a contradiction since the sequence $p(\varphi_m)$ is certainly bounded in $C_P(X)$ [use the characterization e) of the theorem IV.8.1]./ / /

IV.9. $C_P(X;E)$ SPACES OF TYPE (DF) [9] & [17]

THEOREM IV.9.1 [20;58] The space $C_P(X)$ is of type (DF) if and only if for every sequence B_m of P, there is $B \in P$ such that $\cup_{m=1}^{\infty} B_m \subset \bar{B}^{\cup X}$.

In particular, the space $C_c(X;E)$ is of type (DF) if and only if every countable union of compact subsets of X is relatively compact in X.

Proof. The condition is necessary. By the theorem IV.8.1, we know that the sets

$$B_m = \{f \in C_P(X) : \| f \|_X \leqslant m\} \quad (m \in \mathbb{N})$$

constitute a fundamental family of bounded subsets in $C_P(X)$. Let now B_m be a sequence of P. The sets

$$V_m = \{f \in C_P(X) : \| f \|_{B_m} \leqslant 1\} \quad (m \in \mathbb{N})$$

are then absolutely convex closed neighborhoods of 0 in $C_P(X)$ and their intersection V is certainly bornivorous since we have $B_m \subset mV$ for every $m \in \mathbb{N}$. Therefore V is a neighborhood of 0 in $C_P(X)$: there are $B \in P$ and $r > 0$ such that

$$V \supset \{f \in C_P(X) : \| f \|_B \leqslant r\}.$$

Hence the conclusion by the part b) of the proposition I.2.1.

The condition is sufficient. On one hand to prove that $C_P(X)$ has a countable fundamental family of bounded subsets, it is of course enough to establish that every bounded subset of $C_P(X)$ is uniformly bounded on X. If this is not the case there is a bounded sequence f_m

of $C_P(X)$ and a sequence x_m of υX such that $|f_m(x_m)| > m$. As we may of course suppose that the x_m belong to the dense subspace $\cup\{B : B \in P\}$ of υX, we get directly a contradiction since by hypothesis there is $B \in P$ such that $x_m \in \bar{B}^{\upsilon X}$ for every $m \in \mathbb{N}$. On the other hand let V_m be a sequence of absolutely convex closed neighborhoods of 0 in $C_P(X)$ which intersection V is bornivorous. For every $m \in \mathbb{N}$, there are then $B_m \in P$ and $r_m > 0$ such that

$$V_m \supset \{f \in C_P(X) : \|f\|_{B_m} \leq r_m\}$$

and, as V is bornivous, there is $r > 0$ such that

$$V \supset \{f \in C_P(X) : \|f\|_X \leq r\}.$$

Therefore we have

$$V_m \supset \{f \in C_P(X) : \|f\|_{B_m} < r\}, \quad \forall\, m \in \mathbb{N},$$

since for every $f \in C_P(X)$ such that $\|f\|_{B_m} < r$, we have

$$f = \theta_{\|f\|_{B_m}} \circ f + (f - \theta_{\|f\|_{B_m}} \circ f)$$

$$\in (\theta_{\|f\|_{B_m}}/r)V + \varepsilon V_m, \quad \forall\, \varepsilon > 0,$$

hence $f \in V_m$. From this we get

$$V \supset \{f \in C_P(X) : \|f\|_B < r\}$$

if $B \in P$ is such that $\cup_{m=1}^{\infty} B_m \subset \bar{B}^{\upsilon X}$. Hence the conclusion.///

THEOREM IV.9.2.[17]

a) _If_ $C_P(X)$ _and_ E _are_ (DF)-_spaces, then_ $C_P(X;E)$ _is a_ (DF)-_space_.

b) _Conversely if_ $\cup\{\bar{B}^{\upsilon X} : B \in P\}$ _contains one point of_ X _and if_ $C_P(X;E)$ _is a_ (DF)-_space, then_ $C_P(X)$ _and_ E _are_ (DF)-_spaces_.

Proof. a) is established in two steps.

A. Let us first prove that if E is a (DF)-space and if K is a Hausdorff compact space, then $C_c(K;E)$ is a (DF)-space.

By the theorem IV.8.2, we know already that $C_c(K;E)$ has a countable fundamental family of bounded subsets.

Let now V_m be a sequence of absolutely convex closed neighborhoods of 0 in $C_c(K;E)$ which intersection V is bornivorous. Then V is a bornivorous barrel of $C_c(K;E)$ and by the proposition I.6.5, there is a bornivorous barrel T of E such that $C(K;T) \subset V$. For every $m \in \mathbb{N}$, V_m is a neighborhood of 0 in $C_c(K;E)$; this implies the existence of a closed semi-ball b_m centered at 0 in E such that $C(K;b_m) \subset V_m$. Therefore for every $m \in \mathbb{N}$, we have,

$$C[K;\overline{\tfrac{1}{2}(T + b_m)}] = \overline{\tfrac{1}{2}[C(K;T) + C(K;b_m)]} \subset V_m$$

by the part c) of the proposition I.6.3. As the sets $\overline{\frac{1}{2}(T + b_m)}$ are absolutely convex closed neighborhoods of 0 in E and as their intersection T' is bornivorous since it contains $\frac{1}{2}T$, T' is neighborhood of 0 in the (DF)-space E. To conclude it is then enough to note that we have

$$C(K;T') \subset \bigcap_{m=1}^{\infty} C[K;\overline{\tfrac{1}{2}(T + b_m)}] \subset \bigcap_{m=1}^{\infty} V_m = V.$$

B. Let us now consider the general case.

Once again we already know by the theorem IV.8.2 that $C_P(X;E)$ has a countable fundamental family of bounded subsets.

Let now V_m be a sequence of absolutely convex closed neighborhoods of 0 in $C_P(X;E)$ which intersection V is bornivorous. For every $m \in \mathbb{N}$, there is then $B_m \in P$ and a semi-ball b_m centered at 0 in E such that

$$C(B_m;b_m) = \{\varphi \in C(X;E) : \varphi(B_m) \subset b_m\}$$

is contained in V_m. By the theorem IV.9.1, there is $B \in P$ such that $\cup \{B_m : m \in \mathbb{N}\} \subset \bar{B}^{\cup X}$. Every $\varphi \in C_P(X;E)$ such that φ^β vanishes identically on a neighborhood of $\bar{B}^{\cup X}$ in βX belongs of course to anyone of the V_m, hence to V. Therefore the support of V is contained in $\bar{B}^{\cup X}$. Moreover by the part A of this proof, we know that $C_c(\beta X;E)$ is a (DF)-space. Therefore $V \cap CRC_u(X;E)$ is a neighborhood of 0 in $CRC_u(X;E)$: there are $p \in P$ and $r > 0$ such that

$$V \cap CRC_u(X;E) \supset \{\varphi \in CRC_u(X;E) : \sup_{x\in X} p(\varphi(x)) \leqslant r\}$$

hence such that

$$V \cap CRC_u(X;E) \supset P[X;b_p(r)].$$

We get then

$$V \supset \overline{V \cap CRC_u(X;E)}^{C_P(X;E)} \supset \overline{P[X;b_p(r)]}^{C_P(X;E)}$$

hence

$$V \supset C(X;\overline{b_p(r)})$$

by the proposition I.6.2 and even

$$V \supset C[K(V)\ ;\ b_p(< r)]$$

by use of the theorem II.1.4. Hence the conclusion.

b) is known (cf. theorem IV.2.2).///

COROLLARY IV.9.3. *The space* $C_c(X;E)$ *is* (DF) *and quasi-barrelled if and only if* X *is compact and* E *is a* (DF) *and quasi-barrelled space*.

Proof. The condition is necessary. By the theorem IV.2.2, we know that $C_c(X)$ and E are (DF) and quasi-barrelled spaces. To conclude we just have then to prove that this implies that X is compact. If it is not the case, by the corollary III.3.15 of [20], there is a real positive and lower semi-continuous function f on X which is bounded on every compact subset of X and unbounded on X. Hence a contradiction since by theorem IV.9.1, f must be bounded on every countable subset of X.

The sufficiency of the condition is a direct consequence of the previous theorem and of the proposition IV.6.6.///

COROLLARY IV.9.4. *The space* $C_c(X;E)$ *is* (DF), *barrelled and* sq-*complete if and only if* X *is compact and* E *is a* (DF), *quasi-barrelled and* sq-*complete space*.///

COROLLARY IV.9.5. *If* K *is a Hausdorff compact space, if the convention* I.7.1 *holds, if each of the* E_m *is a Banach* (resp. *normed*) *space and if* Ind E_m *is regularly compact, then* $C_c(K;\mathrm{Ind}\ E_m)$ *is barrelled* (resp. *quasi-barrelled*) *and* (DF).

Proof. This is now a direct consequence of the corollary I.7.3.///

CHAPTER V

BOUNDED LINEAR FUNCTIONALS ON $C_P(X;E)$

V.1. REMARKS

Let φ' be a bounded linear functional on $C_P(X;E)$.

By the proposition II.6.3, we know that supp φ' belongs to $\tilde{P}^{\upsilon}$; in particular, supp φ' is a compact subset of υX. If moreover E is metrizable , then there are $p \in P$ and $C > 0$ such that

$$|<\varphi,\varphi'>| \leqslant C\ p_{\text{supp}\ \varphi'}(\varphi), \quad \forall\ \varphi \in C_P(X;E).$$

Therefore by the Singer theorem III.3.4, if E is metrizable and realcompact (resp. if E is metrizable and if X is realcompact), there is a regular countably additive Borel measure $m_{\varphi'}$ on υX with values in E'_s such that $m_{\varphi'}(b) \in C\ b_p^{\Delta}$ for every $b \in \underline{B}(\upsilon X)$, $V_p m_{\varphi'}(\upsilon X) \leqslant C$, $m_{\varphi'}(b) = 0$ for every $b \in \underline{B}(\upsilon X)$ disjoint from supp φ' and

$$<\varphi,\varphi'> = \int \varphi\ dm_{\varphi'}, \quad \forall\ \varphi \in C_P(X;E),$$

this measure being unique.

V.2. MAIN RESULT

DEFINITION V.2.1. The sequence φ_m _is surely converging to_ φ _in_ $C_P(X;E)$ if, for every $B \in P$, there are $m_B \in \mathbb{N}$ and an open subset Ω_B of υX disjoint from $\bar{B}^{\upsilon X}$ and such that supp $\varphi_m^{\upsilon} \subset \Omega_B$ for every $m \geqslant m_B$.

LEMMA V.2.2 [4], [28] _If_ E _is realcompact_ (resp. _if_ X _is realcom-_

pact) and if every surely converging sequence in $C_P(X;E)$ is bounded at the point x_o of υX, then $\varphi_m^{\upsilon}(x_o)$ tends to 0 in E for every null sequence φ_m of $C_P(X;E)$.

Proof. If it is not the case, there is a null sequence φ_m in $C_P(X;E)$, $p \in P$ and $r > 0$ such that $p(\varphi_m^{\upsilon}(x_o)) \geq r$ for every $m \in \mathbb{N}$.

Let us set

$$F_m = \{x \in \upsilon X : \inf_{j \leq m} p(\varphi_j^{\upsilon}(x)) \geq \frac{r}{2}\}, \quad \forall\, m \in \mathbb{N}.$$

Let us establish first by contradiction that for every $B \in P$, there is an integer m_B such that $F_m \cap \bar{B}^{\upsilon X} = \emptyset$ for every $m \geq m_B$. If it is not the case, $F_m \cap \bar{B}^{\upsilon X}$ is a decreasing sequence of non void and compact subspaces of υX. By the Cantor theorem, their intersection is not void; let y be an element of this intersection. On one hand we must have $p(\varphi_m^{\upsilon}(y)) \geq r/2$ for every m. On the other hand y belongs to $\bar{B}^{\upsilon X}$ hence is such that

$$p(\varphi_m^{\upsilon}(y)) \leq p_B(\varphi_m) \to 0.$$

Hence a contradiction.

As the F_m are neighborhoods of x_o, there is a sequence $\psi_m \in C(X;E)$ such that $p(\psi_m^{\upsilon}(x_o)) = m$ and $\operatorname{supp} \psi_m^{\upsilon} \subset (F_m)^{o\upsilon X}$ for every $m \in \mathbb{N}$.

Hence a contradiction since this sequence ψ_m is surely converging to 0 in $C_P(X;E)$ but unbounded at x_o.///

THEOREM V.2.3 [28] If E is metrizable and realcompact (resp. if E is metrizable and if X is realcompact), then every bounded linear functional on $C_P(X;E)$ is sequentially continuous.

Proof. Let φ' be a bounded linear functional on $C_P(X;E)$ and let φ_m be a null sequence in $C_P(X;E)$.

As we have recalled in the previous paragraph, supp φ' is a compact subset of υX and belongs to $\tilde{P}^{\upsilon}$. Moreover there are $p \in P$, $C > 0$ and a unique regular countably additive Borel measure $m_{\varphi'}$ on υX with values in E'_s such that $V_p m_{\varphi'}(\upsilon X) \leq C$, $m_{\varphi'}(b) \in C\, b_p^{\Delta}$ for every $b \in \underline{B}(\upsilon X)$, $m_{\varphi'}(b) = 0$ for every $b \in \underline{B}(\upsilon X)$ disjoint from supp φ' and

$$\langle \varphi, \varphi' \rangle = \int \varphi \, dm_{\varphi'}, \quad \forall \varphi \in C_P(X;E).$$

Therefrom we successively get that

a) we certainly have

$$\left| \int \varphi \, dm_{\varphi'} \right| \leq \int p(\varphi) dV_p m_{\varphi'}, \quad \forall \varphi \in C(\text{supp } \varphi';E),$$

b) supp φ' belongs to $\tilde{P}^{\cup}$ hence

$$\sup_{x \in \text{supp } \varphi'} \sup_{m \in \mathbb{N}} p(\varphi_m^{\cup}(x)) < \infty,$$

c) by use of the preceding lemma

$$\varphi_m^{\cup}(x) \to 0, \quad \forall x \in \text{supp } \varphi',$$

because every surely converging sequence in $C_P(X;E)$ converges in $C_P(X;E)$ hence is uniformly bounded on supp φ'.

The result is then a direct application of the Lebesgue dominated convergence theorem.///

INDEX

REFERENCES

N.B. A reference of the type [20;.] means the reference [.] of [20].

[1] K.-D. BIERSTEDT-R. MEISE, Bemerkungen über Approximationseigenschaft lokalkonvexer Funktionenräume, Math. Ann. 209 (1974), 99-107.

[2] K.-D. BIERSTEDT-R. MEISE-W.H. SUMMERS, A projective description of weighted inductive limits, Trans. Amer. Math. Soc. (to appear)

[3] A. DEFANT - W. GOVAERTS, Tensor products and spaces of vector-valued continuous functions, (preprint).

[4] W. GOVAERTS, Bornological properties of spaces of non-Archimedean valued functions, Bull. Soc. Roy. Sc. Liège, 48 (1979), 413-416.

[5] W. GOVAERTS, Locaal convexe ruimten van continue functies, Ed.ronéotypée, Université de Gand (1980).

[6] W. GOVAERTS, Inductive limits and sequentially continuous linear transformations, (to appear).

[7] A. GROTHENDIECK, Sur les espaces (F) et (DF), Summa Brasil. 3 (1968), 57-123.

[8] R. HOLLSTEIN, ε-Tensorprodukte von Homomorphismen, Habilitationsschrift, Paderborn (1978).

[9] R. HOLLSTEIN, Permanence properties of C(X,E), Manuscripta Math. 38 (1982), 41-58.

[10] A. KATSARAS, On the space C(X;E) with the topology of simple convergence, Math. Ann. 223 (1976), 105-117.

[11] M. LOPEZ PELLICER, A sequential characterization of the ultrabornological spaces C(X), Rev. Acad. Ci. Madrid, 67 (1973), 485-503.

[12] A. MARQUINA-J.M. SANZ SERNA, Barrelledness conditions on $c_o(E)$, Archiv der Math. 31 (1978), 589-596.

[13] J. MENDOZA CASAS, Algunas propriedades de $C_c(X;E)$, Actes VII JMHL, Pub. Mat. UAB 21 (1980), 195-198.

[14] J. MENDOZA CASAS, Algunos resultados sobre los acotados de C(X;E), Actas de les VIII JMHL Coimbra (Portugal) 21 (1980), 195-198.

[15] J. MENDOZA CASAS, Necessary and sufficient conditions for C(X;E) to be barrelled or infrabarrelled, Simon Stevin J. (to appear).

[16] J. MENDOZA CASAS, Barrelledness on $c_o(E)$, (to appear in Arch. Math.)

[17] J. MENDOZA CASAS, Conditions for C(X;E) being barrelled, infrabarrelled or (DF)-space, (to appear).

[18] J. MUJICA, Representation of analytic functionals by vector measures, Vector spaces measures and applications, Proc. of the Dublin Conference, Lecture Notes in Mathematics 645 (1978), 147-161.

[19] J. MUJICA, Spaces of continuous functions with values in an inductive limit, in Functional Analysis, Holomorphy and Approximation Theory, Lecture Notes in Pures and Applied Math. 83 (1983).

[20] J. SCHMETS, Espaces de fonctions continues, Lecture Notes in Mathematics 519 (1976), Springer.

[21] J. SCHMETS, Bornological and ultrabornological C(X;E) spaces, Manuscripta Math. 21 (1977), 117-133.

[22] J. SCHMETS, Spaces of continuous functions, Proc. of the Paderborner Mathematiktagung, Notas de Mat. 27 (1977), North-Holland, 89-104.

[23] J. SCHMETS, Spaces of vector-valued continuous functions, Vector spaces measures and applications, Proc. of the Dublin Conference, Lecture Notes in Mathematics 644 (1978), Springer, 368-377.

[24] J. SCHMETS, Localization properties in spaces of continuous functions, Bull. Soc. Roy. Sc. Liège 46 (1977), 241-244.

[25] J. SCHMETS, Survey on some locally convex properties of the spaces of continuous functions, Bull. Soc. Math. de Belgique XXX (1978), 15-26.

[26] J. SCHMETS, Examples of barrelled C(X;E) spaces, Proc. of the Int. Sem. on Funct. Anal. Holom. and Approx. Prop., Rio de Janeiro 1978, Lecture Notes in Mathematics, 843 (1981), Springer, 561-571.

[27] J. SCHMETS, Bounded linear functionals on $C_p(X)$ are sequentially continuous, Bull. Soc. Roy. Sc. Liège 48 (1979), 158-160.

[28] J. SCHMETS, Bounded linear functionals on $C_p(X;E)$, Bull. Soc. Roy. Sc. Liège 50 (1981), 195-202.

[29] I. SINGER, Sur les applications linéaires intégrales des espaces de fonctions continues, I, Rev. Roumaine Math. Pures et Appl. 4 (1959), 391-401.

Vol. 845: A. Tannenbaum, Invariance and System Theory: Algebraic and Geometric Aspects. X, 161 pages. 1981.

Vol. 846: Ordinary and Partial Differential Equations, Proceedings. Edited by W. N. Everitt and B. D. Sleeman. XIV, 384 pages. 1981.

Vol. 847: U. Koschorke, Vector Fields and Other Vector Bundle Morphisms – A Singularity Approach. IV, 304 pages. 1981.

Vol. 848: Algebra, Carbondale 1980. Proceedings. Ed. by R. K. Amayo. VI, 298 pages. 1981.

Vol. 849: P. Major, Multiple Wiener-Itô Integrals. VII, 127 pages. 1981.

Vol. 850: Séminaire de Probabilités XV. 1979/80. Avec table générale des exposés de 1966/67 à 1978/79. Edited by J. Azéma and M. Yor. IV, 704 pages. 1981.

Vol. 851: Stochastic Integrals. Proceedings, 1980. Edited by D. Williams. IX, 540 pages. 1981.

Vol. 852: L. Schwartz, Geometry and Probability in Banach Spaces. X, 101 pages. 1981.

Vol. 853: N. Boboc, G. Bucur, A. Cornea, Order and Convexity in Potential Theory: H-Cones. IV, 286 pages. 1981.

Vol. 854: Algebraic K-Theory. Evanston 1980. Proceedings. Edited by E. M. Friedlander and M. R. Stein. V, 517 pages. 1981.

Vol. 855: Semigroups. Proceedings 1978. Edited by H. Jürgensen, M. Petrich and H. J. Weinert. V, 221 pages. 1981.

Vol. 856: R. Lascar, Propagation des Singularités des Solutions d'Equations Pseudo-Différentielles à Caractéristiques de Multiplicités Variables. VIII, 237 pages. 1981.

Vol. 857: M. Miyanishi. Non-complete Algebraic Surfaces. XVIII, 244 pages. 1981.

Vol. 858: E. A. Coddington, H. S. V. de Snoo: Regular Boundary Value Problems Associated with Pairs of Ordinary Differential Expressions. V, 225 pages. 1981.

Vol. 859: Logic Year 1979–80. Proceedings. Edited by M. Lerman, J. Schmerl and R. Soare. VIII, 326 pages. 1981.

Vol. 860: Probability in Banach Spaces III. Proceedings, 1980. Edited by A. Beck. VI, 329 pages. 1981.

Vol. 861: Analytical Methods in Probability Theory. Proceedings 1980. Edited by D. Dugué, E. Lukacs, V. K. Rohatgi. X, 183 pages. 1981.

Vol. 862: Algebraic Geometry. Proceedings 1980. Edited by A. Libgober and P. Wagreich. V, 281 pages. 1981.

Vol. 863: Processus Aléatoires à Deux Indices. Proceedings, 1980. Edited by H. Korezlioglu, G. Mazziotto and J. Szpirglas. V, 274 pages. 1981.

Vol. 864: Complex Analysis and Spectral Theory. Proceedings, 1979/80. Edited by V. P. Havin and N. K. Nikol'skii, VI, 480 pages. 1981.

Vol. 865: R. W. Bruggeman, Fourier Coefficients of Automorphic Forms. III, 201 pages. 1981.

Vol. 866: J.-M. Bismut, Mécanique Aléatoire. XVI, 563 pages. 1981.

Vol. 867: Séminaire d'Algèbre Paul Dubreil et Marie-Paule Malliavin. Proceedings, 1980. Edited by M.-P. Malliavin. V, 476 pages. 1981.

Vol. 868: Surfaces Algébriques. Proceedings 1976–78. Edited by J. Giraud, L. Illusie et M. Raynaud. V, 314 pages. 1981.

Vol. 869: A. V. Zelevinsky, Representations of Finite Classical Groups. IV, 184 pages. 1981.

Vol. 870: Shape Theory and Geometric Topology. Proceedings, 1981. Edited by S. Mardešić and J. Segal. V, 265 pages. 1981.

Vol. 871: Continuous Lattices. Proceedings, 1979. Edited by B. Banaschewski and R.-E. Hoffmann. X, 413 pages. 1981.

Vol. 872: Set Theory and Model Theory. Proceedings, 1979. Edited by R. B. Jensen and A. Prestel. V, 174 pages. 1981.

Vol. 873: Constructive Mathematics, Proceedings, 1980. Edited by F. Richman. VII, 347 pages. 1981.

Vol. 874: Abelian Group Theory. Proceedings, 1981. Edited by R. Göbel and E. Walker. XXI, 447 pages. 1981.

Vol. 875: H. Zieschang, Finite Groups of Mapping Classes of Surfaces. VIII, 340 pages. 1981.

Vol. 876: J. P. Bickel, N. El Karoui and M. Yor. Ecole d'Eté de Probabilités de Saint-Flour IX – 1979. Edited by P. L. Hennequin. XI, 280 pages. 1981.

Vol. 877: J. Erven, B.-J. Falkowski, Low Order Cohomology and Applications. VI, 126 pages. 1981.

Vol. 878: Numerical Solution of Nonlinear Equations. Proceedings, 1980. Edited by E. L. Allgower, K. Glashoff, and H.-O. Peitgen. XIV, 440 pages. 1981.

Vol. 879: V. V. Sazonov, Normal Approximation – Some Recent Advances. VII, 105 pages. 1981.

Vol. 880: Non Commutative Harmonic Analysis and Lie Groups. Proceedings, 1980. Edited by J. Carmona and M. Vergne. IV, 553 pages. 1981.

Vol. 881: R. Lutz, M. Goze, Nonstandard Analysis. XIV, 261 pages. 1981.

Vol. 882: Integral Representations and Applications. Proceedings, 1980. Edited by K. Roggenkamp. XII, 479 pages. 1981.

Vol. 883: Cylindric Set Algebras. By L. Henkin, J. D. Monk, A. Tarski, H. Andréka, and I. Németi. VII, 323 pages. 1981.

Vol. 884: Combinatorial Mathematics VIII. Proceedings, 1980. Edited by K. L. McAvaney. XIII, 359 pages. 1981.

Vol. 885: Combinatorics and Graph Theory. Edited by S. B. Rao. Proceedings, 1980. VII, 500 pages. 1981.

Vol. 886: Fixed Point Theory. Proceedings, 1980. Edited by E. Fadell and G. Fournier. XII, 511 pages. 1981.

Vol. 887: F. van Oystaeyen, A. Verschoren, Non-commutative Algebraic Geometry, VI, 404 pages. 1981.

Vol. 888: Padé Approximation and its Applications. Proceedings, 1980. Edited by M. G. de Bruin and H. van Rossum. VI, 383 pages. 1981.

Vol. 889: J. Bourgain, New Classes of $\mathcal{L}^p$-Spaces. V, 143 pages. 1981.

Vol. 890: Model Theory and Arithmetic. Proceedings, 1979/80. Edited by C. Berline, K. McAloon, and J.-P. Ressayre. VI, 306 pages. 1981.

Vol. 891: Logic Symposia, Hakone, 1979, 1980. Proceedings, 1979, 1980. Edited by G. H. Müller, G. Takeuti, and T. Tugué. XI, 394 pages. 1981.

Vol. 892: H. Cajar, Billingsley Dimension in Probability Spaces. III, 106 pages. 1981.

Vol. 893: Geometries and Groups. Proceedings. Edited by M. Aigner and D. Jungnickel. X, 250 pages. 1981.

Vol. 894: Geometry Symposium. Utrecht 1980, Proceedings. Edited by E. Looijenga, D. Siersma, and F. Takens. V, 153 pages. 1981.

Vol. 895: J.A. Hillman, Alexander Ideals of Links. V, 178 pages. 1981.

Vol. 896: B. Angéniol, Familles de Cycles Algébriques – Schéma de Chow. VI, 140 pages. 1981.

Vol. 897: W. Buchholz, S. Feferman, W. Pohlers, W. Sieg, Iterated Inductive Definitions and Subsystems of Analysis: Recent Proof-Theoretical Studies. V, 383 pages. 1981.

Vol. 898: Dynamical Systems and Turbulence, Warwick, 1980. Proceedings. Edited by D. Rand and L.-S. Young. VI, 390 pages. 1981.

Vol. 899: Analytic Number Theory. Proceedings, 1980. Edited by M.I. Knopp. X, 478 pages. 1981.

Vol. 900: P. Deligne, J. S. Milne, A. Ogus, and K.-Y. Shih, Hodge Cycles, Motives, and Shimura Varieties. V, 414 pages. 1982.

Vol. 901: Séminaire Bourbaki vol. 1980/81 Exposés 561–578. III, 299 pages. 1981.

Vol. 902: F. Dumortier, P.R. Rodrigues, and R. Roussarie, Germs of Diffeomorphisms in the Plane. IV, 197 pages. 1981.

Vol. 903: Representations of Algebras. Proceedings, 1980. Edited by M. Auslander and E. Lluis. XV, 371 pages. 1981.

Vol. 904: K. Donner, Extension of Positive Operators and Korovkin Theorems. XII, 182 pages. 1982.

Vol. 905: Differential Geometric Methods in Mathematical Physics. Proceedings, 1980. Edited by H.-D. Doebner, S.J. Andersson, and H.R. Petry. VI, 309 pages. 1982.

Vol. 906: Séminaire de Théorie du Potentiel, Paris, No. 6. Proceedings. Edité par F. Hirsch et G. Mokobodzki. IV, 328 pages. 1982.

Vol. 907: P. Schenzel, Dualisierende Komplexe in der lokalen Algebra und Buchsbaum-Ringe. VII, 161 Seiten. 1982.

Vol. 908: Harmonic Analysis. Proceedings, 1981. Edited by F. Ricci and G. Weiss. V, 325 pages. 1982.

Vol. 909: Numerical Analysis. Proceedings, 1981. Edited by J.P. Hennart. VII, 247 pages. 1982.

Vol. 910: S.S. Abhyankar, Weighted Expansions for Canonical Desingularization. VII, 236 pages. 1982.

Vol. 911: O.G. Jørsboe, L. Mejlbro, The Carleson-Hunt Theorem on Fourier Series. IV, 123 pages. 1982.

Vol. 912: Numerical Analysis. Proceedings, 1981. Edited by G. A. Watson. XIII, 245 pages. 1982.

Vol. 913: O. Tammi, Extremum Problems for Bounded Univalent Functions II. VI, 168 pages. 1982.

Vol. 914: M. L. Warshauer, The Witt Group of Degree k Maps and Asymmetric Inner Product Spaces. IV, 269 pages. 1982.

Vol. 915: Categorical Aspects of Topology and Analysis. Proceedings, 1981. Edited by B. Banaschewski. XI, 385 pages. 1982.

Vol. 916: K.-U. Grusa, Zweidimensionale, interpolierende Lg-Splines und ihre Anwendungen. VIII, 238 Seiten. 1982.

Vol. 917: Brauer Groups in Ring Theory and Algebraic Geometry. Proceedings, 1981. Edited by F. van Oystaeyen and A. Verschoren. VIII, 300 pages. 1982.

Vol. 918: Z. Semadeni, Schauder Bases in Banach Spaces of Continuous Functions. V, 136 pages. 1982.

Vol. 919: Séminaire Pierre Lelong – Henri Skoda (Analyse) Années 1980/81 et Colloque de Wimereux, Mai 1981. Proceedings. Edité par P. Lelong et H. Skoda. VII, 383 pages. 1982.

Vol. 920: Séminaire de Probabilités XVI, 1980/81. Proceedings. Edité par J. Azéma et M. Yor. V, 622 pages. 1982.

Vol. 921: Séminaire de Probabilités XVI, 1980/81. Supplément: Géométrie Différentielle Stochastique. Proceedings. Edité par J. Azéma et M. Yor. III, 285 pages. 1982.

Vol. 922: B. Dacorogna, Weak Continuity and Weak Lower Semicontinuity of Non-Linear Functionals. V, 120 pages. 1982.

Vol. 923: Functional Analysis in Markov Processes. Proceedings, 1981. Edited by M. Fukushima. V, 307 pages. 1982.

Vol. 924: Séminaire d'Algèbre Paul Dubreil et Marie-Paule Malliavin. Proceedings, 1981. Edité par M.-P. Malliavin. V, 461 pages. 1982.

Vol. 925: The Riemann Problem, Complete Integrability and Arithmetic Applications. Proceedings, 1979-1980. Edited by D. Chudnovsky and G. Chudnovsky. VI, 373 pages. 1982.

Vol. 926: Geometric Techniques in Gauge Theories. Proceedings, 1981. Edited by R. Martini and E.M.de Jager. IX, 219 pages. 1982.

Vol. 927: Y. Z. Flicker, The Trace Formula and Base Change for GL (3). XII, 204 pages. 1982.

Vol. 928: Probability Measures on Groups. Proceedings 1981. Edited by H. Heyer. X, 477 pages. 1982.

Vol. 929: Ecole d'Eté de Probabilités de Saint-Flour X – 1980. Proceedings, 1980. Edited by P.L. Hennequin. X, 313 pages. 1982.

Vol. 930: P. Berthelot, L. Breen, et W. Messing, Théorie de Dieudonné Cristalline II. XI, 261 pages. 1982.

Vol. 931: D.M. Arnold, Finite Rank Torsion Free Abelian Groups and Rings. VII, 191 pages. 1982.

Vol. 932: Analytic Theory of Continued Fractions. Proceedings, 1981. Edited by W.B. Jones, W.J. Thron, and H. Waadeland. VI, 240 pages. 1982.

Vol. 933: Lie Algebras and Related Topics. Proceedings, 1981. Edited by D. Winter. VI, 236 pages. 1982.

Vol. 934: M. Sakai, Quadrature Domains. IV, 133 pages. 1982.

Vol. 935: R. Sot, Simple Morphisms in Algebraic Geometry. IV, 146 pages. 1982.

Vol. 936: S.M. Khaleelulla, Counterexamples in Topological Vector Spaces. XXI, 179 pages. 1982.

Vol. 937: E. Combet, Intégrales Exponentielles. VIII, 114 pages. 1982.

Vol. 938: Number Theory. Proceedings, 1981. Edited by K. Alladi. IX, 177 pages. 1982.

Vol. 939: Martingale Theory in Harmonic Analysis and Banach Spaces. Proceedings, 1981. Edited by J.-A. Chao and W.A. Woyczyński. VIII, 225 pages. 1982.

Vol. 940: S. Shelah, Proper Forcing. XXIX, 496 pages. 1982.

Vol. 941: A. Legrand, Homotopie des Espaces de Sections. VII, 132 pages. 1982.

Vol. 942: Theory and Applications of Singular Perturbations. Proceedings, 1981. Edited by W. Eckhaus and E.M. de Jager. V, 363 pages. 1982.

Vol. 943: V. Ancona, G. Tomassini, Modifications Analytiques. IV, 120 pages. 1982.

Vol. 944: Representations of Algebras. Workshop Proceedings, 1980. Edited by M. Auslander and E. Lluis. V, 258 pages. 1982.

Vol. 945: Measure Theory. Oberwolfach 1981, Proceedings. Edited by D. Kölzow and D. Maharam-Stone. XV, 431 pages. 1982.

Vol. 946: N. Spaltenstein, Classes Unipotentes et Sous-groupes de Borel. IX, 259 pages. 1982.

Vol. 947: Algebraic Threefolds. Proceedings, 1981. Edited by A. Conte. VII, 315 pages. 1982.

Vol. 948: Functional Analysis. Proceedings, 1981. Edited by D. Butković, H. Kraljević, and S. Kurepa. X, 239 pages. 1982.

Vol. 949: Harmonic Maps. Proceedings, 1980. Edited by R.J. Knill, M. Kalka and H.C.J. Sealey. V, 158 pages. 1982.

Vol. 950: Complex Analysis. Proceedings, 1980. Edited by J. Eells. IV, 428 pages. 1982.

Vol. 951: Advances in Non-Commutative Ring Theory. Proceedings, 1981. Edited by P.J. Fleury. V, 142 pages. 1982.

Vol. 952: Combinatorial Mathematics IX. Proceedings, 1981. Edited by E. Billington, S. Oates-Williams, and A.P. Street. XI, 443 pages. 1982.

Vol. 953: Iterative Solution of Nonlinear Systems of Equations. Proceedings, 1982. Edited by R. Ansorge, Th. Meis, and W. Törnig. VII, 202 pages. 1982.